Dave
thanks for your support
Joy Carter Minor

Forward
From the Sea

FORWARD FROM THE SEA

JOURNEY OF A BLACK WOMAN OCEANOGRAPHER AND ENGINEER

JOY CARTER MINOR

www.mascotbooks.com

Forward From the Sea

Cover design by Sheldon Sneed

For more information, please contact:
Mascot Books
620 Herndon Parkway #320
Herndon, VA 20170
info@mascotbooks.com

Library of Congress Control Number: 2019901890

CPSIA Code: PRFRE0719A
ISBN-13: 978-1-64307-492-4

Printed in Canada

To my Momma, who taught me
grace, fortitude, and humility.
Until we meet again...

CONTENTS

PROLOGUE

December 4, 1992. It was Friday night. I had just flipped the cassette player to Tracy Chapman's *Fast Car* album to lull my toddler to sleep. I was often criticized for my unconventional parenting. Santa didn't come to our house, neither did the Easter Bunny. I felt I had made too many sacrifices to have my daughter believe in fairy tales. I allowed her to stand at the door and pass out candy to the kids on Halloween only because I wanted her to have a couple of friends in the neighborhood, but I convinced her that dressing up as a different Beanie Baby each year for Halloween was way more fun than trying to figure out each year what to be. A sweat shirt and sweat pants with one of my handmade tags around her neck made her the cutest Valentino Bear and she could wear it to school for Valentine's Day!

When I had learned I was pregnant, I purchased *Before the Mayflower,* Lerone Bennett Jr.'s history of Black America. Every night, I read to my unborn child the history of her ancestors' journey from West Africa, across the Atlantic, and into the twentieth century. During quiet time, before she could walk, I played French language tapes for her. Later, when I returned from long deployments to Africa, she would hear my voice and sit up in her crib to greet me, "*Bonjour*!"

Life is hard. To be able to make hard decisions when she grew up, I wanted my daughter to learn at an early age how to separate fact from fiction. I wanted her to be an informed consumer of goods and information. It was important to me that my daughter learn how to separate what people say and what they do, to be able to take information, even information that's packaged as fact from credible sources to give the illusion that all is well in the world, and form her own critically-derived perspectives. It was important to me that she learn that people are far more important than the beliefs, ideologies, and worldviews that separate us.

When I was sure my daughter was asleep, I turned on the television to the voice of my favorite reporter, Christiane Amanpour, CNN international correspondent reporting from Mogadishu, Somalia. "The unprecedented U.S. military action is now underway." I felt a special kinship with Somalis, having recently spent a considerable amount of time in the Horn of Africa. It was 4:00 a.m. in Somalia. Special Forces, Navy Seals, and Seabee units had approached Somalia, as Amanpour described their approach "from the sea," onto the beach, and towards the airport in Mogadishu. I felt a sudden sense of deja vu as a U.S. Marine described to the awaiting media his frustration that he had come ashore on a rubber boat.

Nostalgia was quickly replaced with sadness as I reminisced and grew more and more concerned for the Somali people with whom I had shared mango tea in the hot East African desert. Two years prior to Operation Restore Hope, which was intended to be a U.S.-led peacekeeping mission in Somalia, I had ended a four-year assignment in the Horn of Africa. In 1987, I had joined the crew of the U.S. Navy operational ship, USNS *Chauvenet* as a civilian to survey the littorals on and near the coasts of Somalia. While deployed on the *Chauvenet,* I had gone ashore, landing on the beaches of Mogadishu with U.S. Navy personnel conning a rigid hull inflatable boat (RHIB). My work, my contributions to charting depths, hazards, navigational aids, and the acoustic properties of Somalia's harbors and approaches, including the littoral zones of Mogadishu, had made possible this operation, and many others like it in East Africa, the Middle East, and other coasts around the world.

I'm thankful for my life's experiences. I earned everything I've ever had or done. Nothing has ever been given to me because I am Black or because I am a woman. Affirmative action, which has been swallowed up by the new Jim Crow, made it possible for my qualifications to be considered among the other qualified applicants. I have been awarded the Armed Forces Civilian Service Medal and the Federal Women's Program Mentor of the Year. I have presented a paper at the American Society of Naval Engineers conference, a proposal that could have mitigated a multibillion dollar loss the Coast Guard suffered in modernizing its aging fleet of surface assets.

Yet, my experiences belie the notion that the American dream is available indiscriminately to anyone who gets an education and works hard. I grew sensitive to the emerging rules that arose from no rules at all, once I arrived. I could smell discomfort when I asked simple questions about upward mobility. I could literally feel the life being sucked out of the room as people in leadership positions began plotting how to usurp all the education and experience that I brought to the table in an attempt to elevate someone with whom they valued more than me. I have been asked to compromise my values to allow others to seek monetary or political gain. My position as a senior analyst has been transitioned to a new hire who came into the job after a couple of years as a nanny, because she demonstrated good management skills. I am often bemused when I hear coworkers talk about being approached at hotel or restaurant jobs and offered positions where they can work up from being an "analyst" to a manager. And it happens. I have never received a random call asking if I wanted a six-figure job just because the caller and I met at a conference a few years prior. I have been assigned as a direct report to others with far less education and experience.

Yes, I have had a seat at the table among the brightest and the best, but an implicit expectation comes with that invitation. It would take several "great opportunities" for me to grasp that I was usually hired to provide the meat and the bones while someone less qualified, with less experience and education, put the skin on my work and got all the accolades for a job well done.

My career path navigating the workforce as a civil servant, a contractor, and as a consultant is unique. Every move I made was strategic and closely bound to maritime strategy. It would take a couple of decades for me to accept two things. First, I believed that people, especially people in positions of power and authority, are inclined to hire and promote those who share a common bond with them. My mistake was in assuming that those common bonds were in education, experience, and associations. Secondly, I believed that by getting as much education as I could, by working ten to twelve hours a day, sometimes more, that my education, experience, associations, and work ethic combined would facilitate my ascent up the corporate ladder. That did not happen. I was met with opposition in almost every position I held, and in most cases, escaping was the only option for me.

My résumé would be much shorter and less interesting had I not encountered blatant racism and sexism. I would never have known that Microsoft Excel had the ability to run scripts in the background had a colleague not modified a spreadsheet I created and changed my data without me knowing it. I would never have learned how to administer UNIX servers had I not been isolated with the intent to have me fail. My curriculum vitae is deep and wide because nothing could break my spirit and quench the fire that propelled me to do more, be more, grasp at everything I wanted, and fight for my rights as well as the rights of others.

I am proud to be one of the few recipients of the Armed Forces Civilian Service Medal for my contributions during the Bosnian War. I endured the despair of the civil war in Somalia that led to the eventual failed state in that country. I left my daughter when she was three months old to report for duty in the Middle East during Operation Desert Storm when Iraq's Saddam Hussein invaded Kuwait. I have been chased by a speed boat through the Straits of Hormuz by locals who had not been informed by their government that my survey team had permission to be in their territorial waters. I've been held in custody for hours by locals in North African and Middle Eastern countries waiting for officials to negotiate on my behalf to determine that my presence was legitimate. I held tight to the command of my superiors, "Whatever happens, don't give up the ship." Yet, my history and the oath I took to "*support and defend the Constitution of the United States against all enemies, foreign and domestic*" has been purged from the annals of military history. So, I get to tell my own story.

I understand that the power of competition in business rests on the credentials of the most educated and experienced employees. I know my résumé shines on a corporate dossier. Minority.

Female. A master of science in engineering from Johns Hopkins University. Thirteen years on Navy ships charting the world's oceans, another 20 years working with ballistic missile defense systems and other highly sophisticated weapons systems. I never doubted my own abilities. I challenged every "no" and questioned every oddity. I dared to beat the odds and focused on disrupting the status quo. I have never been able to control an insatiable desire to see beyond invisible lines of separation. My greatest assets have also been my greatest obstacles. Like a painter, a poet, or a musician can see and feel the beauty around them, I am instinctively drawn to disparity and injustice. I can't turn it off, but I have learned to turn it down. This book takes a journey across the Atlantic Ocean, in the Mediterranean, out of the Red Sea, and into the Indian Ocean with truths of the events and challenges that have shaped (and fueled) my temperament.

Don't you know
Talkin' 'bout a revolution
It sounds like a whisper
Poor people gonna rise up
And get their share
Poor people gonna rise up
And take what's theirs

"TALKIN' BOUT A REVOLUTION"

TRACY CHAPMAN

SCATTERLING OF AFRICA

I waited fearfully on the curb, hoping for Ron to appear, just outside the Djibouti airport. I tried not to make eye contact with the men and women who stared at me as they passed. Their glares intensified my fear. This was not the reception I expected. They were black, but not black like me. Their black was reddish. Their hair was softer. Their eyes were not as white as mine. I could not find any familiarity in these people or this country.

Ron Townsend was an experienced hydrographer. He was smart and cocky. He stood just under five-foot nine, but his confidence was about seven feet tall. I could not understand it then, but Ron had earned his attitude the hard way. He knew his stuff. He was a Senior Scientist, which made him the civilian equivalent to the U.S. Navy Captain on the Navy operational ship, the USNS *Chauvenet*. Ron had graduated from Southern University in Baton Rouge, Louisiana with a degree in mathematics and then broke barriers to pave a way for those of us who followed. He understood the politics of the civil service and the science of ocean surveying. I would learn and he would teach me his way.

Cars pulled into and away from the curb, rattling and roaring, one swerving around the other. I stood with what was left of my nylon luggage. Somehow in transit, the biggest suitcase had been ripped into shreds that now dangled from beneath. I was trying to hold the luggage together to keep my clothes from falling out. I stood on the curb, overcome with anxiety and on the verge of tears, something I had not shed in years.

In that moment, standing on the sacred earth of my ancestors, the moon peeking over the horizon and temperatures well beyond 100 degrees, fear engulfed me. Sights and sounds unknown to me overwhelmed my senses. A shabby white van stopped abruptly at the curb in front of me. I didn't move. Ron hopped out with the swagger of a Black man who was the Senior Scientist on a U.S. Navy ship surveying the Indian Ocean and the coastlines of the Horn of Africa. He had on dark shades and a smirk. He always had a smirk.

"Get in," Ron commanded.

I was eager to get in the shabby white van. The driver spoke bits and spurts of English, but he understood Ron and Ron understood him. He was the local driver Ron paid to take us wherever we needed to go.

"Europa," Ron commanded.

The shabby white van sped out from the curb. I held tightly to my nylon bag that matched the tattered set I had put in the back of the van.

"Did you exchange money at the airport?" Ron directed his commands at me.

My lips didn't move and my throat made no sound, but my facial expression must have said, "What?"

Ron spoke a bit of broken English to the driver, still in a commanding voice. In seconds we were off the main road and speeding down a bumpy dirt road. We passed local men and women carrying, pulling, and dragging loads down the dusty road seemingly in no rush to be anywhere in particular. Boys played in groups and young men squatted in place on the side of the road. Scattered along the roadway were women lying beneath trees, nursing their babies.

We made a left, then a right, another left, and two more rights. We drove further and further away from the sound of older model automobiles honking and wheezing on the unmarked paths in the dirt streets of Djibouti. We finally reached a place where the only sound was the sound of dirt swirling in the air. The van driver pulled up to a stone building where a woman was sitting on a stool. She appeared to be middle-aged, but she looked strong and healthy. She was adorned in a colorful reddish abaya and matching hijab. Her hands were tucked under her clothing. She had on worn-down sandals and her posture quickened when the white van stopped in front of her.

Ron commanded me to sit still, but that command was not necessary. I was in shock. The driver got out first and said something to the woman in a language I didn't understand. He interpreted the woman's reply to Ron.

"Here's the deal," Ron directed his response to me. He proceeded to explain to me what the exchange rate to convert my American dollars to Djibouti shillings would have been at the airport and what the exchange rate would be if I handed my money over to the lady sitting on the stool. I really didn't care. Ron could see I was not ready for this.

"Give me a hundred dollars," Ron commanded.

I complied. Ron spoke to the van driver who spoke to the woman. The woman pulled out a handful of colorful paper, more like wads of colorful paper. As she pulled her hand out from under her clothes, I could see the intricate swirls of henna making loops and twists from her wrist to her fingertips. I also saw the butt of a gun. I drew hot and cold at the same time, in fear. Unknowingly, it was my first glimpse into the dangerous, complex, yet primitive underworld of the clan network that characterizes the Horn of Africa. In the absence of a formal banking system, informal money transfer operators fill the void in a system known as *hawala*. Ron transacted business on my behalf, passed me a handful of the colored paper with numbers on it and commanded the van driver, "Take us to Europa."

By the time we arrived at the whitish-colored Hotel Residence de l'Europe in Djibouti's city center, it had been nearly 30 hours since I had said goodbye to my Momma at the New Orleans airport. I was numb. I was tired. I was afraid. What had I gotten myself into? The streets of Djibouti were far from quiet. I could hear conversations in a foreign tongue and although I couldn't understand the language, the tones were distinguishable. Anger, desperation, business. People were conducting themselves under the light of the moon as though it was the middle of the day. I don't remember checking into the Residence de l'Europe. It is likely Ron conducted that business for me, too.

I cried myself to sleep the first night, the second day, the second night, and the third day. Occasionally, I would peek out of the hotel window at all the strange movement of goods and people, and retreat back to the hotel bed. On the third night, there was a banging at my hotel door.

It was Ron, banging and calling out my name, much like my Daddy had when I overslept and was running late for school. I was dressed. I had gotten dressed each of the past three days, but

I hadn't had the will to face the streets of Djibouti. This was not what I had envisioned. I imagined the quiet, docile kind of Africa where people stared doey-eyed into a camera. No, this was hustle and bustle, hurry up and go, dirt flying, voices bellowing from sun up until sun down, except for the eerie quietness that occurred in the middle of the day, every day. I knew nothing about Muslim prayer times and Ramadan.

I was in awe of the beauty of the Djibouti women. Sarai had been to my room each morning to clean and I had sent her away. She gently turned away, "Yes, ma'am. I will come tomorrow." That she spoke English intrigued me.

Ron's knock and yell at the door jolted me. I opened it to him standing there with his dark glasses and his characteristic smirk.

"Have you even left this room?" Ron barked.

I shook my head.

"Have you eaten?"

Again, I shook my head.

"Let's go. I'll meet you downstairs." He turned and walked away. I gathered my shoulder bag and timidly descended the stairwell, managing a half smile at the young man at the front desk. I went outside into the nighttime streets of Djibouti.

Stepping onto the streets of Djibouti was like stepping into a phantasmagoria. The visual barrenness I had seen from the window of my aircraft as it approached the vast continent had morphed from out of the desert floor, creating human form and moving matter that was all strange and unfamiliar to me. The desert heat seemed to intensify as the sun set. There was scant vegetation amid the hot, arid atmosphere. It seemed inconceivable that human survival was even possible, yet there was an eerie hustle and bustle amid swirls of sand and debris in the wakes of repurposed automobiles.

My gaze instinctively gravitated to the women and children. Destitute women sat by the side of the road, nursing babies while keeping a watchful eye on toddlers; most of them were refugees escaping famine, tribal conflicts and territorial wars. Young boys hustled back and forth, up and down the busy thoroughfare of Djibouti's city center. At times the young boys seemed playful, and yet within moments their romp seemed purposeful as they swooped behind an unsuspecting foreigner to grab a dangling wallet and dash off into the dusty paths. The sound of automobiles barely drowned out the "hubbala" of the natives and the swish-swash of merchants sweeping the sand from in front of their portable stores. A diverse palette of unfamiliar aromas wafted through the Djibouti air and incensed my olfactory organs. The men seemed in a daze, eyes nearly blood red with half smiles, half grimaces on their faces. I would later learn that the men were in a perpetual altered state from a plant-based stimulant called "qat" that numbed them to the mayhem of their existence.

I was emotionally paralyzed, wracked with nervous anxiety among alien spirits. I knew Ron well enough to know that I would not be coddled and cosseted. I was no dummy and I knew how to survive. Yet, the fearlessness and invincibility of my childhood had been reduced to following Ron and doing whatever he did.

I followed Ron across the street from the Residence de l'Europe to a rickety white street cart, which had charred flesh hanging from a metal rod. Ron said something to the thin, dusty, brown man at the cart, who hastily shaved off brown, charred flesh, its juices trickling into a pan below. With no words of indoctrination, Ron handed me the flesh-colored matter wrapped in white paper.

"What is it?" I asked Ron.

"Food" was his reply.

Not having eaten or had anything to drink for three days, I did not argue with common sense. Nothing about the meal wrapped in a flat doughy substance was familiar. The meat inside did not taste like chicken. It did not look like fish and neither did it have the tartness of my Uncle Daddy Willie's deer meat. But it was food and I took a bite. I took a second bite. By the third bite, I felt a sense of relief. The brown meat in the whitish dough became my best friend.

Ron then took me to "fly alley," a narrow strip off the beaten paths of Djibouti streets where local merchants set up shop. As we walked past rows of tables on each side of the alley, the salesmen shouted at us. "Frdee look, for you!" "British or Amedican?" "Amedican Black!" One merchant shouted, "Michael Jackson!" as he held up a stack of cassette tapes. Another merchant wearing a University of North Carolina t-shirt shouted "Michael Jordan!" Luggage, t-shirts, linen, sandals, grooming items, paper, toiletries, you name it and it could be found in "fly alley." It became obvious after swatting continuously, why it was called fly alley. Designer knock-offs seemed to be the hit. Izod shirts had two alligators rather than one. The alligators seemed to be humping one another in the sewn-on patch. United Colors of Benetton was popular in the United States, and there was no short supply of "Benetton" t-shirts in fly alley. Nike's swish and Jordan's signature leap donned the lot of bags, shirts, and shoes.

Ron and I were not the only aliens in Djibouti's fly alley. In fact, there were military personnel and civilians from Italy, France, the United Kingdom, and other African and Middle Eastern nations transacting business with the local merchants. In the cacophony of fly alley, on top of being full from my new-found meal, curiosity and inquisitiveness rescued me from my fear. I became keenly self-aware and my consciousness shifted at that moment. I wanted to know, I needed to know, why in God's name was I in Djibouti? I needed to position myself in this new view of the world and explore why American culture was for sale in the Horn of Africa and why so many people from other nations were here in this seemingly godforsaken place buying a culture that was foreign to the people selling it!

I knew I was in Djibouti to embark on a two-month cruise at sea around the Horn of Africa on a U.S. Navy ship to chart the oceans and harbors of the Gulf of Aden, the Red Sea, and the Indian Ocean. But why? How had time and resources surpassed these people on the streets of Djibouti? Why had such a myopic view of Africa's diversity been supplanted in me despite her being the origin of humanity? The damaging ideologies I had absorbed and my erroneous understanding of the continent from which my ancestors originated, had resulted in me experiencing culture shock that bordered on paranoia. But first, I needed something to drink.

"Where can I get something to drink?" I finally had enough strength to ask Ron a question.

"Don't drink the water."

"Don't buy soda off the street, even if it's in a bottle."

"Only buy a drink in the hotel restaurant or at a bar."

"And don't use the ice, unless you put it in a condom first."

Ron succinctly rattled off the rules in his usual matter-of-fact manner. Ron was all that was grounding me. By the time we walked the length of fly alley, our shoes were covered in sand. I followed Ron into a stone building and just as surreal as the life of Djibouti streets had been to me earlier, I was not expecting to hear the thump and bump of American R&B artist, Levert, belting "*I ain't much on Casanova. Me and Romeo ain't never been friends.*"

Amid low lights, the smell of beer, and smoldering cigarettes, the room was packed with people of all nationalities bumping and grinding to the beat. Men and women were on the dance floor gyrating to "Casanova." Others were secluded against the back wall, sitting on low, faded sofas. The other scientists and U.S. Navy personnel deployed on the *Chauvenet* were either dancing or at the bar. The scene in this place, the Flamingo Bar, was as close to American culture as I had been in the previous week. Even the women were dressed in western clothing. They were beautiful and the men and women were happy to see them.

The one and only time I had been intoxicated was when Alicia Brown and I downed a bottle of MD 20/20 at Mississippi State. After that, I swore off alcohol. On the heels of the previous few traumatic days when I first arrived in Djibouti, I didn't need a buzz to enjoy myself. With a cold can of Coca-Cola in hand, I was able to bask in the pleasures of air conditioning and enjoy observing the seemingly happy people in the Flamingo Bar. The DJ must have played "Casanova" and Eric B and Rakim's "Paid in Full" a hundred times, reinforcing the subliminal messages of love and money that I soon learned was the real source of the club's existence. The beautiful women lived upstairs above the bar. In exchange for shelter, food, and clothes, the women were required to entertain the international visitors who came to Djibouti. The bartender managed the activities of the women in the club. If a dance turned into a conversation, the foreign man or woman (women were enticed as well) would have to buy a minimum of three drinks. If the conversation and alcohol turned into lust, the bartender took care of the transaction to get the couple upstairs to consummate their burning desires. The only limit to male-female or female-female ratio was how much the foreigner could afford.

The meat from the street vendor (I eventually learned that everything that looked like meat was lamb), fly alley, the Coca Cola, Levert and Eric B & Rakim's "Paid in Full" gave me enough life to look forward to the next 30 days at sea on the *Chauvenet.* However, my consciousness had been disturbed by the geographical, cultural, and economic paradoxes of life in Djibouti. As I spent more and more time in Djibouti and Somalia, I developed more and more questions about the local, regional, and international politics that made my assignment to East Africa significant.

Despite the brutal heat, streets littered with refugees, and the primitiveness of existence, Djibouti, the oasis at the Horn of Africa, with its borders at the Gulf of Aden, was as a treasured pearl found in the midst of famine, disease, poverty, and geopolitical conflicts. Its status was a result of European colonialism, intractable tribalism, clan and territorial wars, and corruption that has plagued the region for centuries. In more recent years, Djibouti has been described as an eagle's nest for its strategic geopolitical position and gateway to Africa and the Middle East as well as a magnet for risky investors interested in Djibouti's growing infrastructure. China, France, Japan, and the U.S. have a significant military presence in Djibouti.

Djibouti's economy thrived on the currency of foreigners. A few days before leaving port, I met a young lad who was a member of the French Foreign Legion. He didn't speak English and I could only read a little bit of French. I had been in Djibouti seven days, three days longer than expected due to the ship's supplies being delayed. I had developed a routine of having two meals a day. For breakfast, I ordered *jambon* (French for "ham") from the Le Rift restaurant inside the Residence de l'Europe. The menu at the l'Europe was printed in French. The *jambon* came with two hard, unleavened pieces of *pain* (bread). The Republic of Djibouti is a remnant of French imperialism, which characteristic of colonials, was stolen from the Somali people. Most of the locals are of Somali ethnicity, a proud people who preserved the ancient customs of their culture. A majority of Somali people in Djibouti came seeking refuge from clan wars within the complicated boundaries of their country. Somalis take deep pride in the preservation of their nomadic lifestyle. Because of French, British, and Italian colonialism, French, English, and Italian are dominant languages spoken by the Somali people, even though they tend to speak at least three languages, which includes their clan language.

My second meal of the day was always the lamb that Ron introduced me to on the streets of Djibouti. For two American dollars, I could get two hot, flat pieces of bread stuffed with lamb. However, it took two or three days for my body to cease rejecting the food minutes after I ingested it.

Despite the diverse ethnicities in Djibouti, everything about the country was culturally Islamic. Arabic language and Islamic religion in the semi-presidential republic society was rooted in the Bedouin lifestyle and was the glue that held Djibouti together while its sister-country, Somalia was unraveling under President Siad Barre and his "scientific socialism." I had been briefed prior to my arrival about appropriate attire and taboos, but my actual immersion into the Islamic world overwhelmed me. It would take several trips to Djibouti and Somalia before I shook off the myth of American superiority based solely on the global socio-economic norms in the international system. Globalism was a word and it was clear from the market demand for American culture that technology was having a global effect in the late 1980s and early 1990s as the Cold War was coming to an end. The full effects of globalization had yet to impact the Horn of Africa. The cost to make a telephone call home to the United States was ten dollars per minute.

For four years, I was deployed to Somalia and the littoral zones of East Africa surveying harbors and approaches and identifying aids and hazards to navigation. The coastlands of Djibouti and Somalia at the Red Sea have the busiest sea lanes in the world. Approximately a third of the world's fuel traverses those waters[1]. I adopted Somalia as my African home and the Somali people became my adopted family. During those years, Somalia's President, Siad Barre was becoming increasingly unpopular. His attempt to remove the centuries-old culture of warring clans and corrupt warlords using as a key slogan, "Tribalism divides, Socialism unites," was backfiring due to his own corrupt behaviors and favoring of his clan. After just over 20 years of authoritarian rule, Barre was ousted in a military coup.

I passed up opportunities to travel to Asia because I felt a kindred spirit in East Africa. I have bittersweet memories of Mohamed, the young Somali man who worked at the desk at the Residence de l'Europe. He always pleasantly greeted me as I came in and out of the hotel and we soon became friends. He shared with me the plight of several of his family members killed by a warring clan and how he wanted to one day study in the United States and become a pilot. He convinced me to try other things on the hotel restaurant menu besides *jambon*, as Muslims typically eschew pork. The *jambon* was strictly for tourists. We shared a cup of sherbet, considered such a delicacy to Mohamed that he planned the day in advance. Mohamed accompanied me to fly alley to purchase cassette tapes and helped me select the most popular African music. My favorite was Johnny Clegg and Savuka.

The subject of religion was easy to approach because we established a rapport of mutual respect, me for his Islamic faith and he for my Christian beliefs, though my beliefs at that time were based on very little understanding of the Bible. We decided one day to start at the beginning. Mohamed explained that Somali history was primarily an oral history and this is his understanding of the beginning:

One day the serpent approached Eve and invited her to see something he had seen in the water. As Eve looked into the water, she saw the reflection of a beautiful woman. "Oh, she is so beautiful," Eve said to the serpent. The serpent reached down and extended fruit to Eve. "If you eat this, you will become like this woman." Eve took the fruit and returned to share it with her husband. "Adam, I saw the most beautiful woman today, come let me show you." Adam followed Eve to the water and he beheld a man of mighty stature and good looks. "If you eat this fruit, you will become like this man." And so, together, they ate the fruit and began to make excrement. They were ashamed and hid because they had soiled the garden.

By December 1988, Somalia was engulfed in a civil war. The Somali National Movement (SNM), originally a multiclan rebel group, alleged that Somalia's President, Siad Barre and his government favored other clans over them. Their grievances erupted into clan wars. At the time, I was back at work at the Naval Oceanographic Office (NAVOCEANO) in Mississippi, on my three-month off rotation during which I analyzed depth contours from survey data collected off the coasts of Berbera in Somaliland. One morning I received a call that I will never forget. "Do you know someone named 'Mohamed'?" It was my mother on the other end of the line. "He's in Washington, D.C., and someone is trying to assist him in getting to Mississippi." My heart sank. It was my friend I met in Djibouti, Mohamed. I never understood his real intentions for taking his life's savings and coming to Mississippi. To escape the civil war? For me? I never understood it and it haunts me to this day. Mohamed was my friend and I would never intentionally hurt him, nor lead him on. I had sent him brochures to aviation schools and we had shared some correspondence, but there had never been intimacy or even hand-holding between the two of us. My marriage was almost over, but not because of infidelity, and I did not want to add this to the problems we already had.

Mohamed stayed with my parents for a week. I had a very strained and awkward conversation with him during the time he stayed with my parents. Even now, I get teary-eyed at the loss of a good friend. I realize that I had been ignorant of his culture. I recall briefly talking with Mohamed at my parents' home. He tried explaining something to me about female circumcision, which I did not understand. I realized that religion was not the only things we should have discussed when I was in Somalia. Culture and religion is an absolute for Somali people. Mohamed eventually made a home in Canada and for years he communicated with my parents regularly. I still feel a deep loss at how our friendship ended.

Everything about the Somali people and their lifestyles contrasted my own and what I had learned of Africa. Despite Ron's

Mango Tea with Friends in Bosaso Somalia

warnings, I learned to accept food and drink when it was offered, not daring to miss the opportunity to try something different and careful not to disrespect their way of life. I shared mango juice and tea with a nomadic family in Bosaso in northeast Somalia on the Gulf of Aden. Even desolate places like Fort Jeddid, an abandoned fort, a remnant of past Somali wars and that overlooked the Indian Ocean, became places of solitude and reflection for me.

We did most of our land surveying by the light of the moon when possible and slept by day on cots under a tent. Navy Seabees were responsible for setting up camp and communications. However, the entire survey contingent worked as a team hauling supplies and pitching the tent. Chores were divided among us. It was important to keep the camp area free of food particles because the sand crabs strategically and tactically attacked the camp every night. The crabs somehow knew when the last person in the camp dozed off and would leave trails of claw prints marking their encroachment. The activity of the sand crabs was used to determine who had clean-up duty. The survey team intentionally left a morsel of food for the crab and each person placed a stick indicating which direction they thought the crab would make its approach. The person's marker that was furthest from the place of encroachment was responsible for clean-up the next day.

Assignments that took us from the coastline into the Somali desert were not always pleasant and neither were they comfortable. There was no GPS and we relied on outdated maps, survey markers left by British or Italian soldiers and Somali nomads. Transportation into the desert for land surveys was by helicopter. Navy pilots flew us to the survey area, connected us to a hoist, and lowered us down into the desert sand, careful not to hover too low to avoid getting sand trapped in the helo engine.

I detested desert duty. The sun was brutal. I was a woman stuck in a desert in a tent with Navy Seabees. On one occasion, a sandstorm prevented the helos from returning for three weeks! To pass the time, I would climb to the top of Fort Jeddid after a long, hot, sticky night, singing the hymns I heard Momma sing. One morning, the presence of two hills caught my eye. Not that they had not been there before, but I just happened to notice them that day. The sun was a cool orange, barely peeking over the Indian Ocean, not quite ready to begin the day. I could feel the breeze as it drifted, almost in slow motion, across the camp. The twin hills in the distance were calling my name. I could not judge just how far they were, but I put a couple of oranges in my pocket and set out to explore beyond what I could see in the distance.

I must have walked a couple of hours before reaching the first hill. As I stood there, not sure what I was looking for, I was startled by the presence of a man. He was not much taller than me, but his back was slightly hunched so he had probably been taller at some point in his life. He was dressed in an old tattered Benetton shirt, the characteristic rainbow colors of Benetton unrecognizable, bleached out by the harsh sun. He lived somewhere near or in an opening I could see in the side of the hill from where he emerged. His hair was orange, an effect of malnutrition often seen in the desert. He had very few teeth remaining. For what seemed like minutes, we both stood in complete silence, staring at one another, not knowing what the most appropriate greeting should be. He was probably more used to my kind than I was of his, so he went first.

"British or Amedican?" He spoke clearly in English but his eyes never unlocked from the orange in my hand. By some divine intervention, I offered him my orange, even though when I left camp, I had not anticipated encountering human form. "You speak English?" I asked.

He quickly arrested the orange from my hand, instantly ravaged it, hull and all, while at the same time, turning away from me exclaiming, "With Allah, where there is will, there is way."

On my journey back to camp, I began to feel that the encounter was a pivotal moment in my life. I walked the two hours back to camp along the ocean, occasionally feeling the cool waters trickle through my tennis shoes as they sank in the soft wet sand beneath them, and collected seashells washed ashore. I had a fanny pack with my cassette player tucked inside and bulky headphones over my ears. As I listened to Johnny Clegg and Savuka's "Scatterlings of Africa," I reflected on how the creator and provider had used me in the middle of the Somali desert to provide for a stranger. Neither his gender, beliefs, nor identity mattered at that moment. I did not invite him to my church. I had not stopped to ask him if he was saved, neither did I have him recite the sinner's prayer.

I love the scatterlings of Africa
Each and every one
In their hearts a burning hunger
Beneath the copper sun
Ancient bones from Olduvai
Echoes of the very first cry
"Who made me here and why
Beneath the copper sun?"

"SCATTERLINGS OF AFRICA"
JOHNNY CLEGG

I didn't even ask him who his Daddy was so we could talk for an hour before I could decide if I would give up my orange. I didn't ask him his political affiliation, and I didn't sentence him to hell when he invoked the name of Allah, rather than Jesus. Everything about that encounter shaped my thinking from that day forward. I sensed a complete shift in my consciousness. I now felt driven to explore and negotiate boundaries between religion, politics, and culture. I didn't know how, but I would trust my instincts.

Reality jolted my euphoria several days later when the helicopter returned to pluck the dirty, nasty, sun-burned crew from three weeks in the desert. The sandstorms had made it impossible for them to save us from the brutal elements sooner, although they had come back regularly to drop food and supplies as they hovered from a safe distance above.

I stood impatiently as all the men I had camped with for the past three weeks were hoisted up into the helicopter. When the long line was finally lowered for me, I secured the belt tethered to the long line from the helicopter onto my waist and gave the soldier operating the winch a thumbs up, signaling that I was ready to escape the heat of the Somali desert and to wash three weeks of crud out of my hair and off of my body. I stood with my feet firmly planted in the desert sand beneath the helicopter, holding tightly to the line, waiting to extend my arms and control the twisting that occurs when the hoist is pulled into the helicopter. The helicopter started advancing forward and I could feel my feet being dragged across the desert floor. Initially, I was not alarmed. I assumed the soldier in control of the winch was slow to start my ascent. When it seemed I had been dragged for at least 50 feet, I forced my eyes open and withstood the sand that was nearly sealing my eyes shut to see why the soldier had not started hoisting me into the aircraft.

What I discovered was the crew being entertained by the scene of me being dragged across the desert by a helicopter. They were snapping photos.

MISSISSIPPI REBEL

Travel through Mississippi as far south as you can go and you will stumble upon a unique culture influenced by French and Spanish settlers, the remnants of chattel slavery, and the tragedies of poverty for all ethnicities. Yet, the beauty of the ecosystem, vast stretches of two sleepy rivers, pelicans, cranes, and the occasional alligator is the place I call home in Moss Point, Mississippi. So beautiful to me, my hometown, that I get especially annoyed when people give me the stinky face and ask, "You're from Mississippi?" I grew up where the Pascagoula and Escatawpa Rivers dump into the Gulf of Mexico, 20 minutes east of Biloxi, Mississippi, and 20 minutes west of Mobile, Alabama. Moss Point is far from a sleepy town. My Daddy use to say, "I might be from a small town, but I'm not small time." Briarwood Drive in Moss Point, the street on which I grew up, produced the likes of Eddie Glaude, Princeton University Professor of Religion and African American Studies and author of several books including, *Democracy in Black: How Race Still Enslaves the American Soul;*

Daria Roithmeyer, Professor of Law at the University of Southern California and the author of *Reproducing Racism: How Everyday Choices Lock In White Advantage;* and my younger sister, Janelle Carter Brevard, whose resume includes speechwriter for Secretary of State Condoleeza Rice and Senator Elizabeth Dole.

In less than a two-hour drive, I could be in New Orleans to the west or Pensacola, Florida, to the east. I was told the story of the Pascagoula and Biloxi Indians when I was a very young girl. The Pascagoula Indian Chief fell in love with the Biloxi Indian princess, who was already betrothed to the Biloxi Indian Chief. War ensued between the two tribes. Fearing defeat, the Pascagoula Indians are said to have fearlessly marched into the river to their deaths, singing a song of victory. The portion of the Pascagoula River where it is believed they marched to their deaths is named the "Singing River."

We were accustomed to life where hurricanes frequently ravaged our towns and flooded the streets with disease-infested water that lingered for weeks. No matter how many times I was told to stay out of the water and not swim in the ditches around our house, risking my life to slosh around in the brown, wet muck was too much of a temptation. The swollen waters and the opportunity to gleefully float downstream was worth any punishment I might face for breaking the rules. One day, I drifted to the end of Rose Drive, the horseshoe-shaped street where we lived when I was much younger. A hurricane had come through our community and the streets were flooded. I was sloshing around in the murky, stringy, brown water having the time of my life. In the distance I could see an image walking closer and closer toward me. It looked like my Daddy and he was walking with purpose. His belt was around his neck to keep it from getting wet before he reached me. The water was up to his waist. There was nowhere for me to go other than to walk in Daddy's direction, dragging pine needles and sludge along with me in brown water that was nearly up to my chest. Daddy grabbed me by one arm and whipped my exposed body parts all the way home.

Although I was well aware of the consequences, nothing stopped me from exploring my way into trouble almost every day. Cussin' was an appropriate and effective form of parenting when I grew up, especially for me, because I danced to my own beat. I was the one who pushed all the buttons to see what would happen, walked close to the edge, and pulled on anything that dangled. Cussin', like a good whuppin', meant that your parents loved you, at least that's how my Daddy explained it. A good cussin' from my Daddy always kept me in line. He was always supportive. He never missed a game or a band concert. Most of the time my brother, sister, and I were straight 'A' students. On those occasions when we brought home less than perfect grades, like a 95 on an exam, Daddy would ask, "What happened to the other five points?"

I am accustomed to the innocent, but off-handed compliments of white people who, with good intentions, often complimented my parents on the fact that between the two parents and three children, everybody was smart, or educated, or had college degrees. My mother had a master's degree from the University of Southern Mississippi, my sister has a master in public policy from George Washington University and I seem to never stop collecting degrees. "Ya'll have done good for a Black family," is the usual "compliment" given in a long southern drawl. To which my Daddy would reply, "Hell, we've done good for a white family."

I had my growth spurt early. I was already five foot five and had my first job when I was twelve years old, working legitimately at the Sonic Drive Inn on Market Street. By the time I was thirteen, I was driving an old blue Dodge Challenger to Sonic after school or after basketball practice. I'm not sure how I was able to get a job at that age, or keep a job for that matter. I never gave up basketball. I held my position on the student council and I made straight As. I stayed extremely busy and slept only when near exhaustion. Even after I got home after basketball practice or working at Sonic, I stayed up until the wee hours of the night talking on the phone. My Daddy would charge up the stairs in a rage, grab the phone and put it away where I couldn't find it. He never knew I had an arsenal of phones hidden in the linen closet behind the towels and sheets. As soon as his rage ended and he was sound asleep in his bed downstairs, I would creep to the linen closet and pull out another phone, each one rigged so that no one would hear the phone ring. I had dismantled the phones and wrapped a piece of cloth around the bells. It was the first vibrating phone.

I was not shielded from segregation and racism, neither as a child nor as an adult, but I never bore the burdens of the ignorance of others. When people hurled "nigger" and spit out the bus window at me, I didn't flinch. They were not talking to me. I didn't respond to their vitriol. I am motivated by injustice and disparaging behavior. Daddy raised us to be tough and thick-skinned. He would instigate and taunt us as children and then jokingly say, "Cry nigger, Cry," indoctrinating us into tools of civil disobedience, essential to nonviolent objection. In the Jim Crow South, being desensitized to name calling was essential to keeping us alive. We didn't know how crucial it was for Daddy to acclimate us to the vicious, racist world we would encounter away from the shelter of his watchful eye. When I heard the word, it did not invoke fear in me. That word could not enrage me. My journey would have been completely different had I stalled at the names I was called, the injustices I endured, and the systems intentionally instituted to block my opportunity. Daddy reminded us that we could not change people, but we could change our approach to them. When we went out in public, Daddy reminded us who we were. "You're J.B. Carter's daughter," a reality rooted in integrity, perseverance, courage, and fearlessness. Being called a nigger, or any other name for that matter, never aroused my anger. I knew who I was.

I like to believe that I am a slightly unhealthy, but stable mix of Daddy's aggressiveness and Momma's gentleness. When my sister calls and her children are in the car or when my friend Yvette calls and her husband who is a pastor is in the car, I get greeted with "Don't cuss, you're on speaker phone." Although both "death and life" are in my choleric tongue, I am also mixed with sanguine and phlegmatic temperaments. I have sense enough to surround myself with accountability partners who I can rehearse things I really want to say and have them tell me, "No, you're not going to say that."

My Daddy's momma died giving birth to him. He may have met his father once or twice, and with all of those odds stacked against him, he managed to survive a brutal Jim Crow South, finish high school, then college, and go on to accomplish many "firsts" in his community. Although Daddy only stood about five foot seven, as a young, educated Black man in Mississippi, he was profiled on the hit list of the Mississippi Sovereignty Commission, a state agency that legitimized police harassment of Blacks, particularly those who were active in the civil rights actions of that era.[2] But Daddy was scrappy, quick-witted, smart, and quick with a back hand. One minute Daddy would be dropping one of his signature sayings, like "Once you get it in your head, they can't take it from you." The next minute, and most times without warning, Daddy could slap the taste out of your mouth.

Momma was the penny pincher, stabilizer, peacemaker, confidant, and she represented all that was good in my life. When I missed curfew, Momma would wait in the garage-turned game room to hear the old Dodge pull in the driveway, throw my pajamas out the door and whisper, "Get these on before your Daddy wakes up!" Just as I got in the house, Daddy would stir from his recliner in the family room, the TV still blaring one of those black and white "old bones" shows as we called them like "Murder, She Wrote." Momma would always cover for me, making Daddy think I had gone outside to get a book or something else for school assignments. Although they both graduated from Tougaloo College with degrees in English, truth be told, Momma was way smarter than Daddy, smart enough to let him think he was the smartest. Together, Momma and Daddy modeled for us the responsibility to care for the needs of others.

Every year, we hosted a college recruiting party for young people in the community interested in going to Tougaloo College, my parents' alma mater. Admissions counselors, band directors, athletic directors, coaches, current and former students from Tougaloo would gather at our house. Applications and acceptance letters were drafted in our living room. Football and band scholarships were awarded on the spot.

Momma and Daddy were both community activists, leaders, and active in church. They were both extremely passionate about their fraternities and sororities. Daddy was a member of Omega Psi Phi Fraternity, Inc. and served as the Seventh District Representative for Mississippi, Louisiana, Alabama, Florida, and Georgia during the 1980s. Momma was a member of Delta Sigma Theta Sorority, Inc. and was elected President of the alumni chapter several times. Through their association with the fraternity and sorority they were able to touch the lives of many people, young and old. We traveled as a family to all Omega Psi Phi District Meetings, oftentimes taking less fortunate kids along with us.

We traveled often, but traveling in the Jim Crow South required a lot of planning, a lot of preparation, and a lot of prayer. I have vivid memories of brown paper bags full of Momma's chicken, a loaf of white bread, and grapes Momma had frozen overnight passing from the back seat to the front seat on our road trips to Jackson, Mississippi. But I also remember countless times being pulled over by a highway patrolman and forced to follow him to the home of the "Justice of the Peace," many times just because they could. Watching Daddy say "yes, sir," and "no, sir" to the most vile and demeaning comments was something we never really discussed, but had a profound impact on how I viewed the world. I never understood the dichotomy in Daddy's demeanor when faced with brutal disrespect from the patrolmen. I now know Daddy was keeping us all alive. We followed the patrolman in fear to a house somewhere on a back road to the home of the Justice of the Peace who became judge and jury right then and there. The fine imposed was whatever the Justice of the Peace deemed it to be. We obviously made it out alive, but the emotional trauma remained locked inside each of us. If there was any solace in it all, it was the fact that years later, Daddy would become the Justice Court Administrator and all the highway patrolmen reported to him.

Although *Brown vs. Board of Education* decided in 1954 that segregation was unconstitutional, on September 13, 1962, Mississippi Governor Ross Barnett was heard on the radio and television declaring. "I speak to you now in the moment of our greatest crisis since the War Between the States, we must either submit to the unlawful dictates of the federal government or stand up like men and tell them, never! I submit to you tonight; no school will be integrated in Mississippi while I am your governor!"[3] It was not until 1969 that Mississippi schools integrated. White people viewed integration like it was the apocalypse. Segregation academies were established to prevent white kids from going to school with Black children. One such school in my hometown was Live Oak Academy, so secluded and private, tucked in the middle of woods, that little to no written history exists about the school, at least not publicly. The building is not visible from traveled roads and has been abandoned for many years. The school was once an institution of education for white school-aged children, grade school through high school, whose parents would rather die than send their babies to school with Black children. Ironically, in my own research of Live Oak Academy, the only information I found was one You Tube video of a former student who was told the same secret history I had been told of Live Oak, that it was "ran by the Ku Klux Klan."[4]

All that remains of Live Oak Academy is an abandoned building in the Kreole community of Moss Point and the painful memories of Black and white students breaking racial barriers when Live Oak finally closed for good, forcing many of the white students to be bussed to Moss Point public schools. Another irony of the Live Oak saga is that the students who had no choice but to leave the failing school and matriculate with Black students in Moss Point public schools were far inferior in both academic and athletic ability. As young Black girls, we always knew about the secret societies at Live Oak and wanted to have our own clubs like the "Lockhearts," which we always esteemed from a distance to be for "distinguished" young girls, white girls. When Live Oak students starting trickling into our classrooms, we found the prissy "Lockhearts" to be quite normal under the frilly clothes and flitting mascara.

The year before segregation, Black students had a choice of which schools they could attend. Black teachers like my Momma who had taught at all-Black schools were getting their teaching assignments for the upcoming school year when zoning and bussing would determine the school students should attend. Momma learned she would be going from East Park School to Ed Mayo Junior High School, an all-white school. I'm not sure why Momma forgot to tell me not to talk about her new assignment. I had a reputation for talking too much and telling everything I knew. Nevertheless, I thought it was a good idea to raise my hand in Mrs. Smith's second grade class at Charlotte Hyatt Elementary and announce that my Momma would be going to Ed Mayo to teach English. The horror on Mrs. Smith's crinkled white face and pursed red lips was followed by gasps and blows. I glanced at my best friend Donna Fairley for comfort because even at seven years old, I knew there were racial boundaries, I just chose to ignore them. Without time to escape, Mrs. Smith grabbed me out of my chair, lifted my feet off the floor and shook me violently, screaming "How DARE you say that your Momma is going to take that white woman's job!"

Mrs. Smith may have watered-down my mouth, but she couldn't get in my head or my heart. Even as a seven-year-old child, inside I was affirming myself and my Momma. My Momma was the smartest teacher I had encountered in seven years. Between the efforts of my Momma and Mrs. Lillian Wright at Rainbow Kindergarten, who herself shaped thousands of young black and brown minds, I was an avid reader long before I stepped foot in Mrs. Smith's class. Mrs. Smith had one job and that was to keep my mouth shut. In my innocent mind, I had cause to celebrate. I didn't understand then what had Mrs. Smith all bent out of shape.

I am a mélange of my Daddy's unbridled fearlessness and my Momma's subtle courage. In the third grade, I was bussed to Kreole Elementary. My teacher, Mrs. Barlow, wife of a local Southern Baptist preacher, thought the dress my Momma put on me the first day of school was too short, the dress my Momma drove three and one-half hours to Jackson to buy for the first day of school. Mrs. Barlow took me into the restroom of the trailer that had been converted into a modular classroom in order to accommodate all the students being bussed to the previously all-white school. Why Mrs. Barlow thought it was a good idea to take the hem out of my dress is beyond me. When I had to explain to my mother what happened to my dress, her silent rage went into overdrive. The sky blue dress with a bow held in the center by a safety pin, went flying over my head and catapulted into the kitchen sink for a rigorous hand washing. I saw my Momma's red pin cushion on the side of her bed, as she pulled the needle and thread through that dress, sewing the hem in the exact spot where it had been originally, maybe a little bit higher. This scene repeated every night for the rest of the school week. As soon as I came in from school, the dress was hurled over my head into the sink, except there was no need to hem the dress again. I believe Mrs. Barlow knew that if my Momma had to come to the school, it was not going to be nice.

Daddy was more spontaneous and impulsive. One summer, Daddy had a bright idea to load us in the station wagon and drive from Pascagoula, Mississippi, to Houston, Texas. He had not given thought to the time nor the challenges of driving through Mississippi, Louisiana, and Texas in 1969 without pre-planning our rest stops. By the time we made it to Lake Charles, Louisiana, neither

Momma nor Daddy had the energy to drive any further. It was dark. The brown bag of Momma's fried chicken was gone, and it was obvious we were not going to make it to Houston before daybreak. We pulled into every motel on Highway 90 in Lake Charles, Louisiana, and every one of the desk attendants turned off their amber or red glowing "Vacancy" signs and turned us away. Daddy pulled into an old Gulf Oil gas station and asked the attendant if we could stay in the parking lot long enough to get a nap and be on our way. We were directed to pull to the back side of the gas station. My brother and I wrestled over how we could rest in the back seat without touching one another. Daddy turned the crank to recline his seat and we settled in the station wagon for a few hours rest on a hot, sticky summer night. All that could be heard was crickets and frogs. Daddy's experience as a Black man in Mississippi made him wise enough to leave the windows barely cracked, making the rest stop next to miserable on a southern summer night.

Not long after the car stopped rocking from all of us trying to find a comfortable spot to rest, we all, simultaneously felt eerily as if we were being watched. We opened our eyes to find we were surrounded by white robes and pointed hoods. The sound of them breathing heavily under the robes intensified once they knew we were aware of their presence. A few of the hooded creatures held up signs that read, "Klan meeting tonight. Bring your own Nigger!" Daddy's intuition had prompted him to leave the key in the ignition. Within a moment we were in reverse and spinning away from the Gulf station. That experience and others like it, continues to impact my perspective when I hear people refer to America's urban communities like Chicago, Compton, and Baltimore and even faraway places like Somalia and Sudan as violent, when colonizers and white supremacists were the "architects of violence."[5]

The year our house on Rose Drive flooded for the fifth time in the same year, Daddy put a "For Sale" sign in the front yard while the water was still knee deep in the house. He dragged us off to Regency Woods Apartments with a crazy idea that he could build a house on the white folks' side of town. Their homes and neighborhoods did not flood. By the time the day finally arrived for the house construction on Briarwood Drive to begin, Moss Point Mayor Phillip Watts had ordered the city's garbage trucks to surround our lot. Daddy was not afraid, nor intimidated. He knew somebody, who knew somebody, who knew somebody and by the second semester of my seventh grade school year, we were moving into the house on Briarwood Drive.

Surviving being Black in the Jim Crow South meant to be passive and submissive while suffering in silent pain. This was considered a virtue by white people who mistook the docility and meekness of Black people as genuine mutual love and respect for them, despite being treated less than human in the public square. The smiles and kindness were the only way to survive. The images like those of Emmitt Till, the pre-teen Black boy visiting the Mississippi Delta region from Chicago, who was beaten, tortured, and sunk to the bottom of the Tallahatchie River in 1955, served to secure submission of Black people. Even as a young child, I detested and refused to succumb to being a "good nigger." Langston Hughes penned it best in his poem "Warning."

Negroes,
Sweet and docile,
Meek, humble and kind:
Beware the day
They change their mind!

Wind
In the cotton fields,
Gentle Breeze:
Beware the hour
It uproots trees!

I have vivid memories of being able to fly in my dreams when I was a child. I had a supernatural ability to start flapping my arms harder and harder, I would propel myself into the sky and fly away from danger and trouble. The most difficult flights were when I had to take off from the ground, but if I could make my way to a rooftop, my flight away was effortless. I remember waking up in the middle of the night, hardly able to breathe, because I had flapped my arms so hard trying to escape in my dreams.

I was 15 years old when I accepted the fact that I was not going to literally fly, but that realization did not stop me from trying figuratively. I was accepted to the summer science program for high school students at Tougaloo College, my parents' alma mater. My brother was a sophomore there. Kincheloe Science Hall became my sanctuary, my escape, a place where I could challenge myself and not the world. My new love was learning. As the summer was ending, I walked in the registrar's office with a wild idea, one that would challenge the barriers of college admissions. I walked out of the registrar's office admitted as an early entrance freshman for the fall school year, which meant I would forego my senior year of high school, senior prom, class presidency, and all the accoutrements of being a high school graduate. I would not march with the Moss Point High School senior class, and I would not receive a high school diploma. None of that mattered to me.

I stayed at Tougaloo College long enough to be eligible to transfer to Mississippi State University. I gained practical, marketable experience as a cooperative education student, a program that allowed students to alternate semesters between school and work. I worked at Ingalls Shipbuilding in the Mechanical Engineering Department. My job was to interpret naval architects' drawings and engineering changes and then calculate the differences in fluid flow given the proposed changes to piping designs. I took high steps and long strides in my white hard hat to the gangway and requested permission from the quartermaster to board the United States Navy CG-47 Ticonderoga Class guided missile cruiser. I would traverse port and starboard, down narrow corridors and vertical ladders, down below in the engine room and throughout the galley, locating areas where piping would be changed, making notes of the types of piping designs, being careful to distinguish between sea water, fresh water, and bilge water piping. Using the engineering drawing skills that I acquired in school, I made initial, hand-drawn changes and submitted them to the engineering change review board along with my estimates of fluid flow. I was featured in the shipyard weekly newspaper for my contributions to the team, mostly male engineers, naval architects, and draftsmen.

If I have any regret in life, it is that I gave up on my undergraduate studies in engineering. I transferred to Mississippi University for Women, a short drive from Mississippi State and finished my degree in chemistry. I was also drawn to the "Be all you can be" slogan of the United States Army. I was told by the recruiter that the 1985 Balanced Budget and Emergency Deficit Control Act, known as the Graham-Rudman Act, disallowed the United States military from accepting women into Officer Candidate School who had not taken reserve officer training (ROTC) while in college. Determined to be a soldier, I took the ride to Jackson, Mississippi, with the recruiter and raised my right hand to defend the United States as an enlisted member of the United States Army.

FROM THE SEA –

IN THE BEGINNING

"Our ability to command the seas in areas where we anticipate future operations allows us to resize our naval forces and to concentrate more on capabilities required in the complex operating environment of the 'littoral' or coastlines of the earth. With the demise of the Soviet Union, the free nations of the world claim preeminent control of the seas and ensure freedom of commercial maritime passage. As a result, our national maritime policies can afford to de-emphasize efforts in some naval warfare areas. But the challenge is much more complex than simply reducing our present naval forces. We must structure a fundamentally different naval force to respond to strategic demands, and that new force must be sufficiently flexible and powerful to satisfy enduring national security requirements."

...FROM THE SEA

PREPARING THE NAVAL SERVICE FOR THE 21ST CENTURY
SEPTEMBER 1992
A NEW DIRECTION FOR THE NAVAL SERVICE

The 1980s was an interesting time in both domestic and international politics. More than 40 years of Cold War tensions and hostilities precipitated intimidations and threats from the world's great powers. Since the end of World War II, the United States and the Soviet Union and their allies had kept the world in fear of another major world war. Although propaganda, sanctions, and embargoes prevented the tragedy of another war, the world was keenly focused on the aggression of these two superpowers. The U.S. modernized and expanded its military capabilities in an arms race to meet Soviet aggression. Domestically, affirmative action was the phrase that employers loathed and minorities used as often as it was necessary to obtain employment in jobs minorities were often overqualified for and to ensure equal pay for equal work. Still, Black people were the "last hired, first fired."

Despite the climate in America, especially in the South, I was an eager and willing participant for an all-volunteer military force. I had prepared mentally and physically to fulfill my obligation to the United States Army. So, I was stumped when I received a letter from the United States Navy. I had forgotten about a civil service job I applied for many months earlier. The job was on the Gulf Coast of Mississippi, near my hometown, working for the Naval Oceanographic Office (NAVOCEANO) in Bay St. Louis. I spoke with someone in the human resources office and expressed my interests in the job, but I was on delayed entry for the Army. I was assured that if I really wanted the job, there was a way to obviate the oath I had taken to enlist in the Army.

Thus, on July 6, 1987, as required by Title 5 of the U.S. Code, I took the oath as a GS-5 Physical Scientist at the Naval Oceanographic Office.

"I, Joy Carter, do solemnly swear (or affirm) that I will support and defend the Constitution of the United States against all enemies, foreign and domestic; that I will bear true faith and allegiance to the same; that I take this obligation freely, without any mental reservation or purpose of evasion; and that I will well and faithfully discharge the duties of the office on which I am about to enter. So help me God."

Shortly after starting my civil service career at NAVOCEANO, I received notice from the Department of Defense that I had been enlisted in the Army inactive reserves, meaning I was not entitled to military pay and neither was I obligated to report to basic training to fulfill the earlier oath I had taken to enlist in the Army, unless the President of the United States issued a declaration of war within a period of five years from the notice.

I dared to believe that I was qualified, capable, and more importantly, worthy of a decent job with a critical mission and that I had valuable skills to contribute. What made this job different was the deep feeling of pride and sense of fulfillment I had at the opportunity to contribute to the mission of national defense. Weyerhaeuser Paper Company had extended a more lucrative offer to me, which would have required me to move across the country to Seattle, Washington. Even though the starting salary as a GS-5 civil servant was only $14,000 a year, the oath I took to "*defend the Constitution of the United States*" emanated from the depths of my soul. This was MY country, too! I had finally graduated college and the journey on which I was about to embark as a federal government employee promised that I would be contributing to the defense of America's vital interests.

NAVOCEANO, initially established in 1866 as the United States Hydrographic Office, is located on the site of the John C.

Stennis Space Center in Bay St. Louis, Mississippi.[6] The fact that I was trespassing on anything with John Stennis' name associated with it caused anxiety for my parents. Stennis was a U.S. Senator (D) with a well-known history of supporting racial segregation. Stennis was among the 100 signatories of the *Southern Manifesto* that publicly denounced *Brown vs Board of Education.*[7] He voted against the civil rights and voting rights acts and he was vehemently against the establishment of Martin Luther King's birthday as a national holiday.[8]

None of Stennis' demons or the seeds of his devilment sowed in Sen. Trent Lott could deter me from fearlessly showing up for my government job. When Stennis retired from the U.S. Senate in 1988, Republican Trent Lott succeeded him. The name "Trent Lott" is iconic in Mississippi and especially in Jackson County where I grew up. Despite Mississippi persistently ranking last in everything from access to health care, infant mortality, education, economy, and employment, Southern whites worship the ground Trent Lott walks on. Eighty percent of the students at Trent Lott Middle School in Pascagoula come from low income homes. Ironically, my father is on the airport commission at the Trent Lott International Airport. Trent Lott admitted in a public speech that he was raised to support segregationists' views.[9]

A little-known fact about the activities of the Republican Party during the 1980s is that the same Black people that were once on the Mississippi Sovereignty Commission's hit list became targets of political victimization for the purpose of soliciting the Black vote. The Republican Party's Get Out the Vote strategy was to identify leaders in the Black community and offer them financial incentives to vote and campaign for the Republican Party. My Daddy was one of the people approached and offered financial backing to establish his own business with government contracts set aside just for him, in exchange for his allegiance to the Republican party and to encourage more Black people to vote a Republican ticket. I remember my Daddy adamantly opposing even the notion, but there were friends of his who entertained the proposals. To accept such an arrangement meant betrayal of not only the people in the Black community, but betrayal of civil rights martyrs like Martin Luther King and Medgar Evers, who died fighting for equality. Being a "sell-out," as this was called, was not a legitimate way to level the playing field. To be a sell-out signals a lack of a moral compass, to be someone who doesn't consider consequences and is consumed with self-interests rather than the interests of the Black community. Present-day recruitment of Black leaders is more covert and subtler than when my Daddy was recruited. There have been near scandals based on the previous way Black people were exploited. The truth about Trent Lott's relationship with Hyde Security Services (HSSI) was never made public because every attempt to investigate Trent Lott's involvement with HSSI security services was blocked.[10] [11] HSSI was awarded a NASA contract at Stennis Space Center. Despite the fact that Trent Lott's 71-year-old mother was on the payroll at HSSI, an otherwise good, decent, family man and friend of our family was sentenced to prison for conspiracy to defraud the government.

It wasn't until I took my seat on the first floor in the far northwest corner of Building 1002 that reality sucker punched me. I knew absolutely nothing about oceanography! To my surprise, I was surrounded by people who looked like me, and they ushered me into the field of hydrography, a branch of oceanography that measures physical characteristics of shallow water such as found in coastal areas, lakes, and rivers. Jacques Cousteau, the famous underwater explorer and inventor, was not the only oceanographer and neither did he represent the vast field of oceanography. There were more than 30 hydrographers who greeted me and most of them looked like me.

My supervisor, Lewis "Lew" Robertson, was a GS-14 and head of the Hydrographic Department. Lew was a black man with wire-rimmed glasses, an infectious wide-gapped smile, a Jheri curl, and huge biceps. He walked humbly in tight shirts around both his superiors and his subordinates. Lew, along with several other NAVOCEANO employees, transferred to Mississippi when a deal was negotiated in Congress to move NAVOCEANO from Washington, D.C., to help boost the economy in Mississippi. The close proximity to New Orleans, Louisiana, and the Gulf of Mexico made the Stennis Space Center an ideal location for the Chief of Naval Operations to relocate the Commander, Naval Meteorology and Oceanography Command and its subordinate command, the Naval Oceanographic Office. However, I would later learn that the move to Mississippi was met with skepticism from the beginning. In a case filed in U.S. District Court in 1975 which named Admiral James L. Holloway, III, Chief of Naval Operations and Secretary of Defense Donald Rumsfeld as defendants, the plaintiffs argued that the move was decided on and disproportionately affected a significant number of Black employees. The affected employees were unsuccessful in blocking the move.[12] This would explain why so many Black employees reported to Lew and why Black employees were persistently overlooked for promotion.

The Hydrographic Department was the stepchild of oceanography. During the Cold War, when the U.S. danced with the Soviet Union to be recognized as a hegemon, the Navy focused its efforts on maritime dominance in the deep ocean in anticipation of war with Russia and its allies. Meanwhile, there were storms brewing around the globe. After years of border disputes, Iraq invaded Iran, instigating a war that lasted eight years. Ethnic conflict and civil war erupted with the breakup of Yugoslavia. Libya was suspected of sympathizing with and harboring terrorists. The United Nations Security Council demanded Israel withdraw troops from Lebanon. Not that these regions were unaccustomed to conflict, but the nautical charts necessary for naval and special operations forces to access those areas were out of date.

The career progression at NAVOCEANO for a physical scientist was a two-step promotion process, provided that employee's performance was acceptable. Employees were required to complete two tours of duty at sea annually and successfully complete a training or academic course. A tour of duty involved a two- to three-month stretch. This meant that at the end of the first year, if I met the performance requirements, I would be promoted to a GS-7. If I continued on track as planned, I could be a GS-11 in three years, nearly doubling my salary to a whopping $30,000. After achieving a GS-11, promotions for physical scientists were competitive, meaning most employees were GS-11s in the Hydrographic Department. Advancement was always on my radar from the day I took the oath. My Daddy's words of wisdom had been, "Get all you can and can all you get."

Within the first six months at NAVOCEANO, I departed New Orleans Louis Armstrong airport with three pieces of the five-piece set of wheel-less nylon suitcases my mother provided for me to go on a transcontinental flight to Africa. Characteristically, I felt fearless traveling out of the country, with government orders in hand. I accepted the drudgery of dragging my fragile nylon luggage through customs. I packed adequately for a six-hour layover at London's Heathrow airport, on the route to the Republic of Djibouti at the Horn of Africa. I had gotten instructions from Ronald Townsend, the senior scientist who would meet me once I arrived in the motherland. Ron was also a Black scientist in the Hydrographic Department. Ron had graduated with a degree in physics and was one of only three Black employees in the department promoted beyond a GS-11.

Unfortunately, I was woefully ignorant about where I was going on my first tour to Africa. All I knew about Africa was what I had seen on television. I had not taken the time to grasp the vastness and diversity of Africa as a continent with over 50 sovereign states. I did not seek enough knowledge about the culture, the geography, or the history of the places I would be visiting. When I left America, I was ecstatic about going "home" to the motherland. I had visions of people who looked like me coming out to greet me with open arms, welcoming me home and showering me with handcrafted gifts. I anticipated feelings of nostalgia, as if I had been to this place before. I was sure I would see people who looked like my oldest uncle, Daddy Willie. Going to Africa would be, I thought, like a family reunion of sorts. I transferred from Delta Airlines at Heathrow to an Air France flight bound for Djibouti. My anxiety began to rise as I looked around the aircraft and saw many brown-skinned people adorned in thobes, abayas, and hijabs. The only language I could understand was the accented English of the flight attendant.

The Air France flight from London's Heathrow airport approached Djibouti just as I could see the moon rising from the airplane window. I felt a sense of despair rising in my gut as I viewed the barrenness through my window. Miles of brown, red, hardened, dehydrated terrain, and jagged cracks in the earth from lack of water stretched to eternity. Specks of brown vegetation and swirls of sand disrupted the monotony of Africa's barren land. I sensed the pull of gravity as the plane descended faster and closer. I could see a small section of lights flickering. I had not given thought to what an African airport would look like, nor had my fantasies factored in the ordered chaos that awaited.

It wasn't until I landed at the Djibouti Airport that reality completely disrupted my consciousness. Everything was foreign to me, the smells, the dress, the language, the heat, the cars, the terrain, and the sounds. I was completely overwhelmed and for the first time in my life, fear consumed me.

I had enough wits about me to follow the instructions I had been given before leaving NAVOCEANO. I had both my blue and brown passports, careful not to have my blue passport visible. I knew to only present the blue passport if I needed to go to the U.S. embassy or in specific situations where my life was threatened. I was dressed modestly, but not covered. Djibouti, I had been told, was a majority Muslim country. I had my government orders. I walked quietly through customs and out to the streets of Djibouti. By then, I knew there would not be a family reunion in Africa. No one would be greeting me outside to shower me with tribal gifts. I focused my attention on the information I had been given, that Ronald Townsend would be waiting for me outside. I was afraid.

The first day at sea, as the USNS *Chauvenet* pulled away from the port in Djibouti, escorted by smaller boats with bumpers all around them, I settled into the top rack of my stateroom, staring into the bulkhead above, trying to process the people, places, things, language, clothes, smells, sounds, and feelings I had encountered in the past seven days. Fear resurfaced as I lay, staring straight up, but the sound of water swishing against the ship's hull and the roar of the engines lulled me into a deep sleep.

Ronald Townsend, Senior Scientist on the USNS Chauvenet

I was awakened by a loud "bang, bang, bang" on my stateroom door. "Are you coming to work today?"

"Huh?" I had fallen into a deep, disorienting sleep. But I was pretty sure of the day of the week. "It's Sunday."

It was Ron at my stateroom door. "Get yo' ass up. We work seven days a week. There is no Sunday out here." I was in a strange land on a strange vessel with strange people and my emotional instability and cultural imbalance had just been introduced to seasickness.

I was surprised at how quickly life at sea becomes "normal." Seven days a week, ten to twelve hours per day—out of the rack, up to survey operations, out to the deck, work, eat, exercise, and talk shit all day. It was the work that made this job particularly interesting. Even the monotony of being out to sea for days was disrupted by the nature of the work we did. With a small cadre of scientist, engineers, merchant seaman, and Navy personnel working in concert, we contributed to highly sensitive operations in support of U.S. vital security interests. The ultimate mission was to "chart the world." Existing charts had become outdated; the Cold War was coming to an end and there was a shift from deep water dominance to shallow water. The effects of tides, winds, dredging, construction, and other natural and man-made events prompted a greater need for reliable data along shore lines, coasts, harbors, and their approaches. U.S. Navy personnel, Special Operations Forces, and other clandestine troops frequently accompanied us at sea on assignments that I am proud to have contributed to.

Essential to the morale of those of us on board the nearly 400-foot vessel for 30 to 35 days before returning to shore for supplies was the ability to make contact with family via the ship's radio.

It was the only option for calling home to check on our families. The radio room was the size of a small broom closet; it had a rack of equipment that had switches and lights and a large set of earphones. The radio officer's primary job was to send and receive official communications for the ship's crew, which included uniformed Navy personnel, civil service scientists and engineers, and the merchant seaman who were responsible for the ship's operations. Maritime communications had evolved from Morse code and other ship-to-ship and ship-to-shore communications. The radio officer would connect shipboard communications equipment with a land-based radio officer in the United States via the INMARSAT satellite system. A radio officer somewhere in the United States would receive the radio communication. This was a random connection from the ship to a stateside radio officer who accepted the call as a public service to the men and women around the world on matters of defense and diplomacy.

"Ahhh, this is five one seven four. I would like to patch a call to six zero one, four seven fiver, ahhh five six niner eight. Caller Joy Carter calling Mary Carter. How copy? Over." The ship's radio officer spoke into a handheld box like it was a microphone. When he spoke, he clicked a button on the side of the box. When he wanted a response, he lifted the button.

The radio operator in the states would repeat the number and information and then reply, "Standby. Over."

"Roger, standing by, over." With that, the radio officer removed the bulky vinyl-covered tattered ear muffs from his ears and placed them on my head and secured the muffs over my ears. He instructed me to click the button when I spoke, release it when I was not speaking, and when I completed a sentence, to say, "Over." After a series of clicks, buzzes, and whizzes, I could faintly hear garbled conversations over the radio. Soon my mother's voice rose above the hum.

"Hello!" My mother answered the phone and my heart lit up!

"Ma'am, I have a call by radio from Joy Carter for Mary Carter." After a few exuberant "yesses!" and screams for my Daddy to come into the room, my mother calmed down when the radio officer informed her that the satellite call was ten dollars a minute. He gave her similar instructions as I had received from the ship's radio officer. She had to say "over" at the end of each sentence, signaling him to release the button he was using to communicate with me.

"Hey Momma! Over!"

"Hey Baby! Over!"

"How are you? Over!"

"I'm fine. How are you? Over"

"I'm fine. Over."

"It's so nice to hear your voice. Over."

"OK! I love you. Over!"

"I love you too! Over."

"Bye. Over."

"Bye. Over"

Although the actual conversation was only one minute, the $50 was money well spent.

Life at sea had a disorienting, stressful culture of its own and not because the crew was unwelcoming nor were they unfriendly. In fact, there was a weird sort of camaraderie on the surface. The merchant marines contracted by the Military Sealift Command cooked and served in the ship's mess hall. The handful of civilian scientists onboard worked alongside uniformed crew in planning the oceanographic survey missions, deploying hydrographic survey launches (HSLs), and rigid hull inflatable boats (RHIBS) for transporting people and equipment ashore. Living, eating, and socializing among enlisted and officers was separate but superficially and ceremoniously respectful.

Women were accommodated on Navy operational ships, like the oceanographic survey platform I was deployed on, long before the Navy allowed women on combat ships. Call it naiveté, but I did not anticipate experiencing hardships at sea based on my race or gender, and neither did I expect special favors either. I was expected to carry my weight, just like my male counterparts. When we launched the smaller boats for a day's work on the beach, I was required to lift and load all the survey gear through the hot African desert. I tied the boat off at the pier in whatever knot I could finagle. If the boat ran aground, I was duly expected to jump out of the boat and push or pull it into the currents. At the end of a long hot day installing tide gauges, conducting triangulation surveys, or an even longer and brutally hot day of reconnaissance for a future mission, my clothes were just as wet, soggy, and sandy as the men's on the survey team. Special treatment was never expected and never received. So when I was startled awake on the *Chauvenet* in the wee hours of the morning, while the moon was still high over the Indian Ocean, to the loudest, most threatening bang on my stateroom door, I immediately went into a threat posture assuming the worst: That the ship was under a hostile takeover.

Whoever was on the other side of my stateroom door was relentless in trying to open it and I could tell they were not alone. I could hear commotion and chaos as others were being forcibly removed from their rooms and bullied down the ship's corridors. I could see from the porthole of my stateroom that we were still sailing. The ping of the depth sounders was still bouncing off the bulkhead below my stateroom. There was so much mayhem at my door, I stood behind it fully expecting the perpetrators to eventually kick the door in. There was no way to peep through the door to see who was violently tugging at the knob. I feared for my life, but I was surprisingly not paralyzed by my fear.

Subconsciously, I had confidence that I would be protected. I had even more confidence in the fact that I would be safe just off the coasts of Somalia and Kenya. I had become a friend of Somali people and they had extended only graciousness toward me. I did not view them, nor treat them in an "uncivilized" manner as my shipmates had, especially the young Navy sailors. I had been appalled and insulted at the presumption of young sailors that Africans were barbaric and uncivilized. Too often the youngest, most inexperienced, poorest sailors treated the Somali people with total disrespect. Too often I intervened when young sailors assumed the Somali merchants could not count and often tried to take merchandise they had not paid for. The condescending tone in which they spoke to the pastoral people on the occasions where we had to get permission to survey on territory where nomadic families had settled, drew fire inside of me. I often found myself having to negotiate on behalf of, and unbeknownst to the U.S. Navy, to get permission to survey when the sailors conducted themselves in undiplomatic ways. My favorite line for the ignorance of the young sailors was to remind them that Uncle Sam did not live in Somalia.

I had an uncanny confidence that I could reason with whoever was on the other side of my door and spare myself harm or even more importantly, save my life. But it soon became clear that the voices outside my door were American, only I could not discern what they could possibly want with me at such an ungodly hour.

"You dirty, slimy pollywog, Joy Carter!" Someone was yelling at my door.

Things turned weird at that moment, because whoever was at my door, knew my name and it was a familiar voice. I yelled for them to tell me what they wanted.

"You've been called to stand before the royal court of Shellbacks!"

Honestly, my first thought was, "This is some white folks' stuff right here," but I vaguely remembered something being discussed the previous evening about crossing the equator. I recalled a shipmate asking if I wanted to become a Shellback, sort of an initiation for crossing the equator. We had been sailing across the Indian Ocean headed to Mombasa in order to make a switch of the survey team. I would be going home at the next in-port. I remember saying something about my willingness to participate in the Shellback initiation. I had pledged a sorority and I never gave much thought to the Shellback initiation being this dramatic, so the conversation discussed in the wardroom had slipped my mind.

"I'M NOT DOING IT!" I yelled to the rousers outside my stateroom door. I heard the thunder of their boots scurry down the corridor, amid continuous wrestling, jostling and insults to other "dirty stinking pollywogs." Although I could hear they had left me alone, I was not content that this would be the last time I heard from them.

Things did quiet down, leaving me a couple of more hours before I had to report to the survey operation room for a 12-hour shift. Although it wasn't a peaceful sleep. I managed spurts of 15- or 20-minute naps, in between watching the flicker of light as boots walked down the corridor in front of my stateroom to see if the feet stopped at my door. I really didn't need my alarm clock because my stateroom was directly across the corridor from the entrance to the engine room. The engineers came in and out of the door to do their 6:00 a.m. checks, which was more effective at waking me than my alarm clock that was often drowned out by the sound of the engine humming, the depth sounders pinging, and the waves of the ocean slamming against the bulkhead.

I emerged from my stateroom very cautiously, not quite sure who or what would be lurking behind a staircase or compartment door or hatch. There was a short walk forward and one ladder up to the mess hall. I figured if I made it that far without incident, all was well. As I approached the bottom rung, I stepped aside to allow James Causey to walk past. James was one of the other NAVO scientists and a man of few words. A southern gentlemen who "took (his) supper" at the same time every day and finished it in less than five minutes. During supper he didn't talk with his long, drawn-out southern drawl. In fact, James never looked up when he was taking his supper. When his head went down to eat, he ate. Everyone at the table could hear that James enjoyed his supper and no one dared tell the six-foot-something, burly James that he was eating too loud. But as I stepped aside to let James come down, he was almost at the bottom rung. I turned my head sideways to see if James really had his underwear on the outside of his overalls.

"James? What? Where? What's going on?" James never had much to say, but not in a rude kind of way. James was there for the money and nothing else. When he did engage in conversation, it was usually very matter of fact and straight to the point, unless the conversation involved a math or physics problem, then James had quite a lot to talk about. James acknowledged my inquiry by responding, "Nuttin'," and he walked past me to go to his stateroom to sleep until his next watch later that day.

I continued on into the wardroom where the ship's Commanding Officer (CO) sat with a couple other shipmates. The ship steward, Tony, didn't acknowledge my presence as he normally did so that I could place my order for breakfast. I pretended to be reading the menu, which rarely changed very much from morning to morning. Eggs and some sort of meat were always on the breakfast menu, which was typed by the chief steward the previous night and placed in a plastic binder. The chief steward moved the cream of wheat, oatmeal, grits, and potatoes around on the menu each day, depending on how close we were to the next in-port for restoring supplies. Since we were a day or two away from the in-port at Nairobi, the choice was cream of wheat. But it was obvious that Tony was intentionally ignoring me because someone else walked in and he immediately took their order. Even when I summoned him by name, he carried on as if I had not spoken a word. I was confused by his behavior and had dismissed the dirty looks I was getting from the CO across the table from me. Besides, I was friends with all the guys in the steward department. I had learned after my first deployment that the long time between dinner and breakfast required friends with access to the galley. But not only that, my parents instilled in us to treat everyone with respect, equal respect including my colleagues, host nation partners, and locals in the areas where I was deployed. Most of the time at sea, there some was semblance of reciprocity and inclusiveness. We conducted a myriad of operations at sea, near the shore, and on land, in regions where intractable conflict resulted in war and natural disasters led to famine. My upbringing was not the only reason I treated all my shipmates equally, but I was constantly aware that the microcosmic community onboard the ship could easily be disrupted by instruments of war just across the border in the areas where we conducted survey operations.

One day I was approached by a young, obviously privileged merchant seaman officer with a request, "We would appreciate it if you would not fraternize with the hired help." I'm not sure who the "we" was who sent this young man as a representative. Admittedly, the backgrounds of some of the stewards employed by the Military Sealift Command were questionable. There was a death by overdose on one of my first three deployments, but I still subscribed to mutual respect of all persons and I had learned how to be selective on when and how to respond to actions I perceived to be divisive, motivated by hatred, and which served no more than to debase and disparage groups of people. The request from the merchant seaman was one of the times when I chose to channel my Daddy's wit and use it appropriately with the only language sailors understand. "Unless you own this bitch, we're all hired help," and I walked away, never to be chastised by that person again.

I couldn't understand why Tony was ignoring me. I only had a few minutes for breakfast anyway. I still had the gaze of the CO on me when he broke the silence. "What are you, like a GS-5 or something?" The CO asked me a question. I still had the menu in my hand and I peeked over it and respectfully replied, "Yes sir."

"Well, if you ever want to be a GS-7, I suggest you comply and report to the call of the Shellbacks." The CO's tone and demeanor were non-negotiable. I was given time to return to my stateroom and turn my blue Navy-issued overalls inside out. My underwear had to be outside of my clothes, turned inside out as well. Once my attire complied with the orders, I returned to the wardroom and Tony plopped a plate of green eggs and ham in front of me, which I did not eat.

For most of the day while I monitored survey equipment for depth contours, floating or moored objects on the bottom of the ocean, and regular checks for satellites signals, the day was pretty normal except for the fact that I was dressed inappropriately. I didn't even bother going in the wardroom for lunch. I had a stash of cheese crackers my Momma sent in a care box. The whole pollywog thing had me spooked. I knew things would only get weirder and they did.

I was looking forward to the end of my 12-hour shift. Based on earlier events that day, I didn't expect a decent meal but it had been a long day and I took a chance on going down for dinner. I knew something was awry before I even got into the mess hall. I could hear much of the same commotion I heard the night before. Yelling and screaming "dirty, stinking pollywog!" I had no choice. I was really "stuck between the devil and the deep blue sea" as my Momma used to say. I strapped on an extra dose of courage and entered the room.

Eight or nine of my shipmates were all lined up on the floor, including James, which was really weird. I was shoved down on the floor with them and told to get on all fours. Bowls of what looked like table scraps were placed in front of us and we were commanded to eat. Like the rest of the dirty stinking pollywogs, I hesitated to devour the disgusting food, but we received incentive from the lashes on our hind parts with pieces of rubber, like the kind used on the ship's fire hoses. If James was bothered by the meal, it didn't show. He was the first to finish.

When we had all eaten our table scraps, we were commanded to crawl up to the flight deck, three flights up. It seemed like hours for us to get to the top deck because crawling on all fours while being whipped with the hoses caused many pit stops on the way up. James was the last in the lineup of dirty sticking pollywogs and even though he had very little to say, when James needed a break, he took it. Over half of us would be at the top of a flight, another three or four of us would be half way up the flight, and James would be taking a break on the landing below. The more we stopped, the more lashes we got. When James finally decided he could crawl his 250-pound frame up to the flight deck we were engulfed in seawater from the fire hoses. The Shellbacks punished our laziness and stinky-ness by dousing us with sea water from a high pressure hose, just enough to sting, but not enough to cause bodily harm.

Finally, on the flight deck, we painfully crawled across the hard steel ship's deck, being whipped with the rubber hoses, and doused with seawater. But the hazing was just getting started. The night sky was pitch black but the flight deck was well lit. The pollywogs were separated into four groups and taken to each of the four corners of the flight deck. In the first corner I was taken to, I was commanded to get inside a large piece of leather. It looked like it had at one time covered a piece of ship furnishing, but it was cut in the shape of a fish. It was almost as long as my 5'6" frame and almost as wide. The Shellbacks told me my assignment was to go inside the belly of the whale and retrieve the fish, a dead fish. I was coaxed inside the leather "fish," which was full of galley garbage seemingly put through a meat grinder. It smelled like raw sewage. Again, negotiating with the Shellbacks at that point was a non-issue. The darkness of the belly of the "whale" and the smell of sewage made it impossible to see the fish, but I wallowed around in the muck until I felt something big enough and not yet ground up. I stuck my hand out of the whale's belly with the dead smelly fish tightly in my grasp. I could hear cheers of triumph as the

Shellbacks hosed me down when I emerged from the belly of the whale. They marched behind me as I crawled to the next station.

An entire toilet bowl sat on the flight deck. My orders—retrieve the "treasures" in the toilet bowl without using my hands. Figure that one out. I bobbed for hotdogs that were floating around in a toilet bowl filled with chocolate liquid. There was even more exuberance as the shellbacks cheered me on to the next station. I could also hear the cheers as the other pollywogs conquered their tasks. Each victory was followed by a gush of seawater from the hoses and a few more lashes on my hind parts.

I crawled to the next station. By then, I was a hopeless captive. I had lost track of the day of the week and the time of day. There was nowhere to run and there was no horizon to be seen in the

Pollywogs get no respect

distance, not even the occasional lights of fishermen at sea. If the ship were still sailing, I could not tell. I was focused on survival. The initiation was unlike anything I had ever experienced pledging Delta Sigma Theta.

I was commanded by the Shellbacks to kneel in front of a chair where the big, fat, hairy deckhand was sitting with his legs wide open and his shirt off. His 50-inch hairy belly was covered in a jelly-like substance and he had a cherry stuck in the middle of his naval. I had to grab the deckhand by his knees while retrieving the cherry from his naval with my mouth. Victory was rewarded with cheers, lashes, and seawater by the Shellbacks.

The time between stations got longer as the night dragged on. Some pollywogs took longer than others to find the fish in the belly of the whale or to retrieve all the treasures from the toilet. And of course, James set the pace for everybody. At one point, James stopped for a cigarette break. We were all tired, our bodies ached, and we embodied what one would imagine a dirty, stinking pollywog to be like.

I was directed to open my mouth for the "vaccination" required for all the "diseases" pollywogs carry. There was not a distinguishable flavor in the concoction that was pushed down my throat with a syringe. At that point, I was a prisoner of war. From what I could see, there were no more stations for the Shellbacks to drag me to. My hind parts were given a few more lashes as I crawled to the formation of pollywogs that were assembling on the flight deck as each of us conquered the commands of the Shellbacks. I have no idea how long the pollywogs were lined up on the flight deck facing the helicopter hanger. We were allowed to stand for the first time since the evening dinner. We stood shivering from all the water we had been doused with. The heat of Africa could not warm us up and the sun had not yet appeared. As if we had not endured enough, the Shellbacks harassed and interrogated us while pouring a mixture of catsup and eggs in our hair, only for them to hose us down again with seawater.

Suddenly the ship's emergency alarm sounded for only a few seconds and a voice came over the ship's intercom. "Here ye! Here ye! This is Davy Jones!" I had no idea what was happening and I was too exhausted to care. I don't remember what was spoken by this Davy Jones person, but I do know that I was somehow declared a Shellback and the entire ship erupted into shouting and celebration. Almost simultaneous with the celebration, it was announced that we had crossed the latitude and longitude on the map signaling we had crossed the equator.

As each of us was congratulated and allowed to go to our staterooms for the first time in 24 hours, we were given the most beautifully designed Shellback certificates, one for the wall and another for our wallets. I understand if I ever cross the equator again, I must have the wallet certificate to prove I am a Shellback.

Over the course of three or four years, I spent almost half my life on the coasts of the Indian Ocean. At the end of each tour, I would spend a few days in Djibouti or Mombasa before returning home to reunite with my family. At the end of one long tour, after packing to leave the ship, I thought I would visit my friends in their staterooms, most of whom were male merchant seaman. I walked to the staterooms on the lower deck where most of the deckhands and stewards slept. The guys usually hung out together in one room playing cards, drinking, talking shit and telling lies. That's how I was able to piece together how the Flamingo Bar in Djibouti operated. It was not unusual for me to open the door while knocking at the same time because you could hear their deep voices laughing and joking before reaching their rooms. I noticed it was unusually quiet this particular day, especially since we were in port. Usually, there would be a lot more excitement and noise. At least that's how things had been each time we pulled into port at Djibouti. I didn't waste too much time trying to figure out what was different. I opened the door and there were my friends, four or five guys snorting cocaine. I was more afraid than they were. One or two of them looked up at me and smiled. The others never even noticed I was there or that I had quietly pulled the door and walked away. We had only been at the port in Mombasa for an hour and somehow cocaine was already on the ship. That was my first and last time seeing cocaine.

I was more than relieved to disembark the ship once the watch officer lowered the stairs that allowed me to descend the gangway. I set out to make my way to the Hilton Hotel in downtown Mombasa, but not before being greeted by all the Kenyan men and women on the pier waiting to sell something to us Americans, whether legal or illegal, tangible or abstract. There was a long line of Kenyans waiting with glossy business cards and flyers that promoted nightclubs to disco. And then there were the shady characters, men and women who could not explicitly describe what they were selling unless you turned a corner away from the ship.

Despite my Momma's warnings, I have always been a loner. I don't like what I call, "group therapy," my term for what happens when you travel in groups and have to decide by consensus what to do and where to go. I liked traveling alone, deciding for myself when and where to go. I had not gone far from the ship when I was overcome by an unnerving, uncanny feeling that something was amiss in Kenya. I couldn't put my finger on it, but I trusted my instincts. Rather than take advantage of the one perk that civilians had on the ship that the military did not, I chose to stay on the ship for the two nights before my Delta Airlines flight to New Orleans, rather than check into the Hilton. I went back to my stateroom, found a good book and settled in for the evening. The past few hours had been physically and emotionally draining. It was easy to settle into the familiar than to navigate the busy streets of Kenya.

The ship was unusually quiet that first night in Mombasa. Tony couldn't wait for his shift to end so he could join the rest of the crew frolicking around the city. We laughed about the previous day when Tony had been directed not to speak to me as long as I was a pollywog. As soon as Tony's shift was over, he rushed out into the nighttime. I walked around the ship making small talk with the sailors who were on watch that night. The ship was tied to the pier, so very few merchant seaman were left onboard. The engine was shut down and everyone who was not on watch was granted leave. The long line outside the purser's office had long been cleared from seamen making draws from their checks; the medic had dispensed condoms to all who needed them. The crew had previously experienced a scare where 80 percent of the ship's crew, including officers, had some sort of sexual contact with each other. Everybody had to be tested for venereal diseases.

I was standing on the deck, listening to the sound of life in Africa. I had no regrets from choosing a quieter alternative that night. I noticed a car pull up to the gangway. The driver, a taxi driver, got out of the driver's seat and walked to the back door. I thought it was unusual that the taxi driver would open the door for someone coming to board the ship. But the passenger did not emerge from the vehicle. He remained motionless in the back seat with his head lolled back. The taxi driver rolled him by his shoulders out of his car and onto the asphalt at the foot of the gangway. The watch officer called on his radio for help and they rushed to the end of the gangway to carry the young man to the enlisted sailor's lounge. It was Tony. I asked the watch officer what was wrong with him. "Too much to drink," he replied. I had seen enough and decided my stateroom was the safest and sanest place for me to be.

I was only in my stateroom for a few minutes when I looked at my watch. It was just after 8:00 p.m. Tony and I had spent a good bit of an hour laughing and talking starting at around 5:30 when I went up for chow until he cleaned up the galley and rushed out into the night. He could not have left the ship until 7:00 p.m. But by 8:00 he was passed out on a sofa in the lounge? Things did not add up for me. I rushed back to check on Tony. He had not moved from where the sailors left him. For my own satisfaction, I took his pulse. I thought I felt a pulse, so I left Tony, even though he still had not opened his eyes or moved a muscle. I did not smell alcohol.

I went back to my stateroom and watched the time. At 9:00 p.m., I went back to check on Tony again. The medic was there. The CO and the Chief Engineer were there. They were all hovered over Tony. I heard the medic say, "Time of death"... and he quoted several times, zulu time, local time... I faded out. Tony had died from an overdose of uncut cocaine he bought on the streets of Mombasa. His body was placed in the ship's freezer until arrangements could be made to send him back to his family.

My drama in Kenya seemed to never end. There was something

going on and there would be no American carriers flying out of Kenya for three weeks. We were told we could not fly one of the many Air France or Emirates Air flights out of the country. Michael Smith, Tom Crew, and I were told to check in with the American embassy and then get a hotel room and wait until an American carrier could get us out of Africa, a minimum of three weeks.

I was initially devastated at the thought of another three weeks in Africa. After watching Tony die and stuffed into the freezer, I was traumatized. The first couple of days, I shopped with the street vendors of Mombasa. I met a young National Geographic photo journalist. I must have shown too much excitement over his photos of African pygmies because he started stalking me. I came in from a day of shopping and had a nice pair of gold and ruby earrings waiting for me at the front desk. It wasn't until Michael Smith pretended to be my boyfriend that I was able to stop him from waiting around the hotel for me to show up.

After a couple of days in Mombasa, I was caught in a massive wave of tourists and locals trading with merchants and conducting business. The streets of Kenya were as condensed and bustling as New York's Times Square, mostly black and brown people, but just like any other metropolitan city, the city center was a melting pot of world cultures. I was long past the shock of Africa. Although I had been disappointed after not finding a resemblance among the Somali people, their acceptance and care for me led to a kinship I intend to carry into eternity. They were my relatives from some ancient time passed. But in Mombasa, I saw faces of people with whom I was sure I had close genealogical ties. I was in awe of Mother Africa. Nothing in American media had prepared me for the contemporary and economic progressiveness of Mombasa in Kenya. When I did lower my head from gazing at the tall buildings,

I saw someone who had an uncanny resemblance to my cousin Joe. Our eyes locked—me at him and him at me. The pace of our strides became slower and slower as we approached each other among many other people weaving in and around us.

I believe we both spoke at the same time.

"Joy?"

"Joe?"

Yes, it was my cousin. We are related by marriage but in Mississippi a cousin is a cousin. What were the odds that I would encounter a real Mississippi relative on the streets of Mombasa? Joe Davis was with a team recruiting athletes in Kenya. He invited me to tag along with him to a boxing match. I still erupt into a fit of laughter every time I think about that night. I had my fifteen minutes of fame that night. The ring girl carried a sign with my name on it and the announcer enthusiastically called my name, "Introducing Joy Carter from the United States of America!" The crowd belted out an uproar of thunderous applause as if I were a world-renown celebrity!

I found a "saloon" (hair salon) where I could get my hair braided after convincing the Ethiopian stylist that I was not from any particular tribe that I could call by name. Tom Crew, a fellow oceanographer, called his grandmother in Michigan and she connected us with Dr. Ngundo, a Kenyan professor who had stayed with Tom's grandmother when he was an undergraduate student. The Ngundo's invited me to stay in their home in the Parklands, where Mount Kilimanjaro was visible in the distance from their expansive estate and where I spent most of the time that I was stranded in Kenya.

The civil war in Somalia and the emergence of extremists groups resulted in that country drifting into anarchy. There is a well published account of how the U.S. exited Somalia:

As 1990 came to a close in Somalia, the embassy headed by Ambassador James Bishop had 37 American staffers, having been through successive draw downs as civil war and mob violence spread through the country. Though the embassy attempted to conduct basic operations, gunfire was heard outside the compound and some of it was aimed inside. Asserting his ambassadorial authority, Bishop refused to allow the Marines – either those assigned to the embassy or those who came later aboard the rescue helicopters – to return fire from the Somalis. Given the extreme volatility of the mobs, that policy proved a very sound one.

The final rescue occurred on January 5, 1991, when two CH-53s from the USS Guam were sent to evacuate the remaining Americans and nationals of 30 countries who had sought refuge in the compound. As the last helicopters left Mogadishu, intruders came over the walls, killing several Somali embassy workers and, probably inadvertently, destroying the warehouse containing food supplies that might have brought some relief to embassy Foreign Service Nationals (FSNs) left behind. At the end of the evacuation, 281 people, including eight ambassadors, 61 Americans and 39 Soviets, had been brought to safety.[13]

And then there is a less known account of the U.S. involvement in Somalia. Michael Shanklin, a former major and Vietnam veteran in the Marines who served 13 years in the CIA, was deputy chief of station in Mogadishu in 1990. Shanklin grew up in the Watts area of Los Angeles, most notoriously known for the Watts Riots, when African Americans rebelled against police brutality. Insurgents in Somalia were mobilized and armed to end the brutal 20-year dictatorship of Mohamed Siad Barre. It was Shanklin's ability to navigate the culture of the Somali people that played a significant role in much of the success attributed to James Bishop in evacuating the embassy in Mogadishu.[14]

Cultural apathy and ignorance in the U.S. has proven repeatedly to be the weakest link in both domestic and foreign affairs. Being cognizant and respectful of Islamic culture, clan dynamics, and politics was not as much of a challenge for me as it was for the rest of the crew. Once leaving Djibouti, I was asked to choose between "African" and "African American." I was happy just being "Black" before then. For years after that, the word "African" was scribbled over in pencil on my tourist passport. My official passport was tucked away and never presented unless I had to visit a U.S. embassy. Sailors, both enlisted and commissioned carried a badge of superiority everywhere we went. Although I was not much older than them, I became the mother hen for young sailors who assumed they could disrespect the culture of the host country. I reminded them of what it was like being marginalized and devalued in our own country.

Once, while docked in Bahrain and settling in to a good night's rest, I was awakened to loud knocking on my hotel room door. Naturally, we were always on high alert following the Gulf War and terrorist attacks around the world. This time it was the hotel manager. All U.S. personnel were being evicted from the hotel and the ship's crew was banned from ever returning to Bahrain. It seems two sailors had mistaken the intent of an Arab businessman at the bar. Somehow the businessman was found hanging upside down by his feet, unclothed, out of a hotel window. Thankfully, the business man was still alive or we would probably still be in Bahrain today.

FROM THE SEA –

THE BEGINNING OF THE END

U.S. foreign policy and military strategy kept syncopation with technological advances, modernization, and globalization during the 1990s. Scholars still debate the good, the bad, and the ugly of these changes as nuclear weapons were increasingly finding their way into the hands of rogue actors. Further, the U.S. failed miserably at peacekeeping and democratization in the Middle East and Africa. U.S. military forces were more focused on joint exercises both between the services and with their alliances abroad. The Goldwater-Nichols Reorganization Act of 1986 charged the Chairman, Joint Chiefs of Staff, with the responsibility to assist the President and the Secretary of Defense in providing strategic direction for the Armed Forces. General Colin Powell, Chairman Joint Chiefs of Staff issued a new National Military Strategy built on four key foundations—Strategic Deterrence and Defense, Forward Presence, Crisis Response, and Reconstitution. The Berlin Wall had fallen and there was this fallacy that the world would be a much better place if all nations adopted the American form of democracy. Countries in Eastern Europe were abandoning a weakened Soviet Union and communist ideologies in exchange for more democratic forms of government. Unified forces led by the U.S. were able to halt Iraq's invasion of its neighbors in the Gulf.[15] There was little room to celebrate these victories, as the world began to grapple with other threats in air, space, and land, at sea and in cyberspace. Throughout my career, I was engaged in all of these threat areas.

My first port of entry in the Middle East was in Dubai, one of the seven emirates in the United Arab Emirates. When I arrived in Dubai the first time, the emirate people were just recovering from economic losses following the Gulf War. Investors were developing Dubai's infrastructure. Gold-adorned lights on paved roads, modern telecommunications, and the most sophisticated irrigation systems one could imagine, existed in Dubai, which is within the Arabian desert. I was assigned to the USNS McDonnell on my first tour in the Middle East. The McDonnell was among the aging fleet of oceanographic platforms. NAVOCEANO deployed a team to re-survey areas where nautical charts were considered outdated in published notices to mariners. The team was also among the first to integrate and deploy early global positioning system (GPS) technology as well as multibeam bathymetry and sonar imaging. Using side scan sonar technology, I assisted uniformed officers and enlisted sailors in locating U.S. military aircraft that had been downed during the Gulf War. I was proud to be on that team of scientist, not only because of my contribution to support my country but because the USNS McDonnell was built in Moss Point, Mississippi where I grew up.

For most of the 1990s, I had an integral part in transitioning emerging technology in the maritime domain. The basic elements of hydrographic surveying consisted of collecting water depths that were correlated with a navigational position, as well as other environmental observations like the speed of sound in water. Sound speed profiles are a factor of temperature and conductivity (i.e. the amount of salt in the water). As underwater imaging technology matured, I became proficient at launching side scan sonars off the stern of the hydrographic survey launches and getting images of things that sat on the ocean floor or those things that may be slowly moving at the bottom of the ocean. Side scan images were primarily useful in mine countermeasure operations to inform the U.S. Navy of fixed or floating mines in international waters. Occasionally, the survey team worked along with Special Operations Forces (SOF) and Navy Seals to develop charts to assist them in rescues ashore or to locate items on the bottom of the ocean that were critical to security. Somewhere in the lost records of my years as an oceanographer, is a letter of commendation from the U.S. Navy when I assisted a team of Navy Seals in plotting sea surface temperatures.

Having GPS satellites in orbit revolutionized activities at sea. After years of utilizing crude radio and terrestrial positioning apparatus, GPS satellites were both a blessing and a curse to the oceanographer, particularly the hydrographer who surveyed shallow water, coasts, harbors, and approaches where the water depths were shallow and subject to constant changes. Shipboard GPS receivers required a "fix" on at least three satellites in order for the ship's navigation to be considered on-station. Watching the GPS receiver became a full time job. Whenever the receiver indicated that there was a lock on less than three satellites, another crew member had to mark the navigation-bathymetry (depth) data as speculative to be factored out later in post processing. We spent more time repeating surveys in areas where GPS dropped below three satellites than we did using the old radio positioning methods. Lost satellites meant dead reckoning and often times we had no idea how far we had drifted from the survey area.

Kinematic and differential GPS was the response to surveying the littorals, or those areas on and near the shoreline that were most vulnerable to errors due to lost satellites or other environmental disturbances. My days consisted of being launched into the sea in a rubber boat and driving the boat on the shores of Bahrain and Jabel Ali in Dubai. Usually there were two surveyors assigned to a kinematic GPS survey. I walked for hours in temperatures exceeding 100 degrees, with a GPS receiver in a backpack on my back and a satellite receiver in my hand along the shores of Middle Eastern coastlines. The only thoughts that made the assignment bearable was my infant daughter at home with my parents and shopping in the gold souks of Dubai once the ship took us back to port!

Kinematic Survey on the Arabian Gulf

One of my responsibilities while deployed at sea was to stand aboard a 35-foot boat outfitted with sophisticated survey equipment when it was launched from the ship. The boat was referred to as a hydrographic survey launch (HSL). HSLs were attached to the survey vessel by a davit, a crane that suspended the boat out from the ship and over the water. The boat, while still attached to the davit, would swing like a pendulum from side to side. In the HSL, one member of the survey crew would stand forward and another crew member would stand aft, both outfitted with thick gloves and a hardhat. Once the boat was a safe distance away from the ship, but still suspended and swinging in midair with the crew struggling to stand post, the crane would slowly lower it and us into the choppy waters below. After the boat was in the water, with seas often thrusting the small boat into the mothership, those of us on the boat's deck would give the signal to the davit operator. Our job was to release the enormous hook from the wrecking ball that secured the boat to the davit, having to risk the danger of being beaten up by waves all the while enduring the rocking and rolling of the boat and releasing the hook from the davit.

Survey operations were conducted in many places where there were ethnic wars and civil unrest. The only thing standing between us and the local fisherman or host country government officials who spoke Arabic, Swahili, Somali, or the hundreds of tribal languages in the Middle East and Africa was a piece of paper inside of a plastic binder. That was our "get-out-of-jail-free card." The survey team was advised to hand the paper over if we were ever intercepted while working at sea or on the shore installing or retrieving tide gages. Unfortunately, rarely did the local people know why we were there. Inevitably, we would be stopped and escorted to the local rudimentary police station or military headquarters.

It was considered rude to decline the offers of mango tea. I also had to pretend the men were not staring at me as if they had seen a ghost; a woman, in the desert, on a U.S. navy ship, with men!

We were required to fly an American flag off the stern of the Hydrographic Survey Launches. That alone was enough to draw attention to us. We learned that in most cases if we stocked up on extra sandwiches from the stewards before going out for a twelve-hour survey day, that we had sufficient resources, in addition to the get-out-of-jail-free paper, to spend an hour or two held captive before being released to complete our day. There were no cell phones, just handheld radios. Oh, and there were radios installed on the HSLs. The Captain on the ship knew when we were being followed and when we were being detained. "Don't give up the ship," was the only advice we were given. And "Do you have your 'get-out-of-jail-free' paper." I didn't give a damn about that ship or that paper. I used common damn sense to keep myself alive!

Two- or three-month deployments at sea broke up the monotony of office duties. When I was at NAVOCEANO, I would steal away to the Maury Oceanographic Library where I could expand my knowledge of acoustics, bathymetry, sonar, and other oceanographic instrumentation. One day, I stumbled upon an article, *From the Sea*, while perusing the aisles of the library journals. The perception that federal employees' performance was not properly recognized was mirrored at NAVOCEANO in 1992 when I encountered this article. I was a young woman, working in a predominantly male-dominated work environment, although there was a good representation of women being deployed on Navy ships. *From the Sea*, subtitled, "PREPARING THE NAVAL SERVICE FOR THE 21ST CENTURY, A NEW DIRECTION FOR THE NAVAL SERVICE," laid out a new Navy and Marine

Corps strategy from then Secretary of the Navy, Sean O'Keefe, Admiral Frank B. Kelso II, Chief of Naval Operations, and General C. E. Mundy, Jr., Commandant of the Marine Corps. I was really just doing what was natural for me, reading. But this article provoked many feelings, feelings of hope, feelings of opportunity, feelings that I had a valuable skill. *From the Sea* was about shifting the nation's focus to the littorals, which I interpreted to mean that my value, my worth as a scientist, would become even more significant and would open opportunities for those of us in the Hydrographic Department. I was eager to learn the next steps. How would NAVOCEANO align its workforce with this new strategic directive? What role would I play in this military and strategic posture in the littorals?

I didn't understand it all. I had not been exposed to how Washington works. I knew nothing about the President's national security policy or any precursors to military and defense strategies. Within this article, there was one paragraph that stated the Navy would be concentrating more on "capabilities required in the complex operating environment of the 'littorals' or coastlines of the earth." With the demise of the Soviet Union, the nation was shifting its focus from deep ocean capabilities to the shallow zones and coastal areas.

Well, this was it! For the previous five years, I had been deployed to East Africa and the Middle East dozens of times to chart the harbors and approaches using single-beam and multibeam depth sounders to chart the depth of coastal zones. I had mastered deployment of conductivity, depth, and temperature (CDT) probes and the analysis of sound speed profiles to determine locations in the ocean suitable for mine countermeasures and other subsurface warfare. I was considered an expert in analyzing sound speed profiles in the littorals. I interpreted sonar images looking for aids and hazards to navigation. I had been dropped in the Somali Desert by helicopter, before the existence of the global positioning system (GPS), and conducted triangulation surveys. I could install and monitor tide gauges. I had charted coastlines in the United Arab Emirates with a differential GPS receiver on my back and a satellite dish in my hand. I had been accosted by the locals on several occasions while driving a rigid hull inflatable boat in either the Arabian Gulf or Indian Ocean. I had charted acoustic profilers for safe navigation in the Adriatic Sea. The only diplomatic reinforcement I had was a piece of paper written in Arabic and instructions not to surrender the boat. I knew the littorals. I just needed to know who to talk to. This would be easy, or so I thought.

For a brief moment, I thought about the deep ocean scientists working upstairs in our building. They knew the deep ocean, but the scientist downstairs, all we knew was hydrography and we knew we were the stepchildren of oceanography. And here it was, by mandate of the Secretary of the Navy, this was our time.

It has taken 20 years for me to be able to discuss the obstacles and character assaults I faced when I decided to challenge the institutional racism that affected the African-American employees at NAVOCEANO who were college-educated, excellent scientists, and experts in the littorals. Systematically, one-by-one, we were directed to train the deep ocean experts who quickly became the shallow-water experts and were promoted to all the positions of management and oversight. By then, I was a GS-11, the unofficial limit for Black employees at NAVOCEANO. There were a couple of exceptions, as most institutions tend to cherry-pick one minority, satisfy a quota, and dismiss any other minorities that would otherwise be qualified for promotion.

I published a few articles in the *Command* newsletter and I also published a professional paper in the *Marine Geodesy Journal*, "Telemetering Hydrographic Data," which predated GPS and the Internet. The article examined the use of INMARSAT satellite for transferring data from the area of operation to the Defense Mapping Agency (now National Geospatial Intelligence Agency) for near-real-time publication of hazards and dangerous conditions at sea. Previously, the standard procedure was to hand-carry hard drives with massive amounts of data to be processed months later and distributed sometimes years after data collection. Existing technology also enabled the ability to include warnings in daily SITREPS, Situation Reports, transmitted to the U.S. via the radio officer.

After being denied numerous promotions for GS-12 hydrographic positions by determination that I was "not qualified," my natural instincts to challenge went into overdrive. But first, I wanted to make sure that I exhausted all reasonable, less aggressive options. I drafted a letter to the Commanding Officer, Larry Warrenfeltz, the Executive Officer, Timothy McGee, the Technical Director, Landry Bernard, and the Human Resource Manager, Mr. Delgado. I described my enthusiasm after reading *From the Sea* and my disappointment in what I perceived to be disparity in promotions and opportunities, and the disproportionate number of Black employees in the Hydrographic Department.

Mr. Delgado was the first to reach out to me. We had a previously established rapport, as I regularly volunteered for events that promoted awareness for minorities and women. I had been awarded the Federal Women's Program Mentor of the Year, I was the Vice Chair for the Black Employee Program, and I was an officer for the Stennis Chapter of Blacks in Government. I was surprised that Mr. Delgado had neither heard of *From the Sea* nor read the pivotal article. My letter had been very specific and detailed. I literally developed a strategic plan for NAVOCEANO, suggesting that cross-training was necessary to have a well-balanced, well-prepared workforce capable of deploying in both deep water and littoral zones. I recommended a reorganization of the agency that would establish deployment teams with diverse skill sets to work together at sea, leveraging the skill sets of all available scientists. I suggested that the ethnic and gender mix of the workforce be examined to see if disparity was factual.

Not only had Mr. Delgado not read *From the Sea*, even as the Human Resource Director, he had not realized that more than half of all Black employees at the agency were in the same department working for the same Black supervisor and that only three of those employees, Lew Robertson, Ron Townsend, and Stan Harvey, had been promoted beyond a GS 11. On top of that, he did not seem to realize the last of those promotions was nearly ten years in the past! I accept the fact that I am sensitive to disparity. I see it, feel, it, and I discern it when I walk into a situation. I also sensed that pressure had been pushed down to Mr. Delgado to assuage my concerns and extinguish what could be a firestorm for this federal government agency, especially since "littorals" was the Navy buzzword of the day.

People are paid to craft strategic responses to sensitive subjects. I was told the agency would "look into the matter" and get back to me. I continued to apply for positions; many of the position descriptions repeated my job qualifications verbatim. I moved from the "not qualified" candidates to the "qualified, but not selected" category. In the meantime, I researched as much as I could about workplace discrimination. There was very little information available and minimal recourse for federal government employees to file grievances. The federal employees' union seemed flummoxed in their response to my inquiries. After several months of contemplating, applying, and being rejecting, I concluded that I had stewed in the pot long enough and it was time to elevate the level of pressure.

Because it has taken me more than 20 years to resurrect the traumatic sequence of events that unfolded after I filed a class action law suit, the timeline of what occurred is blurred in my mind. What I remember most is what I felt and how I was shunned, even by the employees who stood to gain from the class action case. I was the Class Agent, representing more than 20 employees' in a discrimination case against the agency. I remember including statistics on the Black employees' dates of hire and promotions compared to those of white employees. At that time, I was unaware of the previous case filed on behalf of the Black employees who were disproportionately affected by NAVOCEANO's relocation to Mississippi from Washington, D.C. I was only aware of the obvious. Black employees at NAVOCEANO were disparately affected by the agency's adjustment to the navy's vision for a post-Cold War world. I quoted *From the Sea* heavily, and made what I thought was a strong case for the agency to consider the strategic restructuring of office personnel to align with the Chief of Naval Operations' strategic vision.

I remember getting a call from Mr. Delgado, who asked me to visit him in his office. He was upbeat and had a smile on his face, as if what he was about to tell me would make us both happy. He informed me the good news. I was to be promoted to a GS-12. However, because I was no longer a member of the class of employees named in the class action suit, I no longer had commonality with the class and by virtue of my promotion, the class action suit would be voided. I treated that conversation with Mr. Delgado like a contract to sell my soul. Rather than walk away, elated to get a promotion, essentially being a "sell-out," I focused on the purpose of the grievance. There was nothing in the letter I wrote to the Commanding Officer or the class action complaint about promoting me. This was a classic case of appeasing and pacifying, basically a way to sweep the real issues under the rug. What was more important to me was to address the systemic disparity that had existed since the inception of the agency. I was more concerned about the future of all minorities, not merely a promotion for myself. I could have filed a personal grievance with the agency had my concerns been self-centered. Even with other Black employees reminding me that they were "making more money than (they) had ever made in (their) lives," and that I should "be satisfied having a good government job and let it go," the essence of this challenge was not about good salaries and job security. It was about fairness and equality and a level playing field.

The cycle to eliminate commonality by promoting the class agent would repeat several times before no one else had the courage to accept the role of class agent in order to keep the case alive. By then, the agency was well aware that change was necessary. The agency quietly reorganized and restructured. The demographics of the agency were addressed and disparities were proven. Diverse survey teams were established and the agency facilitated cross training. In fact, the agency initiated the Total Quality Management concept to document existing processes and operations and improve the agency's overall effectiveness. I was an integral part of that effort.

There was an unforeseen, unwelcomed consequence of the grievance I filed, despite the hint of progress as the agency reorganized and attempted to remove the disparity in its employment practices. By daring to disrupt the status quo, I had essentially put a target on my back. I had not anticipated the resentment and alienation I would incur. I realized there was no room for error or slack in my performance, in attendance, in my associations, how I dressed, where I went, and what I said. It is a lesson I would carry with me for the rest of my life. However, that experience forced me to undertake a personal evaluation and I challenged any hint of complacency that had surfaced within myself, just as I had challenged the institutional status quo that had oppressed a class of people.

After accepting the GS-12 position, I compiled a strategic plan for myself. My near-term goal was to apply and become accepted to attend the Royal Navy Hydrographic Long Course, the only graduate-level course of its kind that satisfied the International Hydrographic Office (IHO) requirements for a Class-A Hydrographer. These were credentials required for NAVOCEANO to certify the quality and accuracy of survey data and nautical charts transmitted to the National Geospatial Intelligence Agency. Michael Quitman Smith was the only other employee that NAVOCEANO had sponsored to attend the rigorous practicum and program of classroom and field study at HMS Drake in the United Kingdom. The program was taught by Royal Navy officers and universities in the United Kingdom. The course was certified by the IHO and administered by the University of Plymouth. The first year I applied for the highly selective program, I was told I was "not qualified." I continued to take engineering courses at the University of New Orleans (UNO), as I had done every year since I became a federal government employee. The UNO courses satisfied my performance requirements for promotion but also satisfied a yearning

I had, since I deeply regretted not completing my engineering degree at Mississippi State.

The second year I applied for the Royal Navy Long Course, I was told the agency would not accept applications for the next course. I applied again the third year. I'm sure there was fear at the agency that if they repeatedly rejected my application, I might escalate my concerns. The agency made an offer that in hindsight, I probably should have declined. My Daddy had a saying, "Do you want what you want, when you get what you want, and discover the price you have to pay." I was selected to attend the six-month course, which would require temporary duty transfer to England. I informed the agency that my then six-year-old daughter would be accompanying me.

HMS Drake is located in southwest England in Plymouth, where the Mayflower set sail for the Americas. I was so driven to be great that I accepted the selection to attend the Long Course with the caveat that the office was only willing to pay half of the Joint Travel Regulations (JTR) per diem. There was no allowance for dependents. I still believed that the sky was the limit and in the meritocratic belief embedded in the illusive American dream, that if I worked hard, my opportunities would be limitless. I still had faith that education and hard work were the keys to success. My Daddy used to say, "Once you get it in your head, they can't take it from you."

Michael Smith had been the only person in the United States to attend the Long Course. I would be the second. I knew the course would be challenging. Michael had not done well in the course, yet he had returned to a promotion as the designated chief hydrographer authorized to sign nautical charts, as required by the IHO.

To say I struggled in England is an understatement. The first challenge was finances. The government-issued Diners Club Card I used for official travel strangely had its spending limit reduced. I discovered this when the rental car agent at Heathrow Airport declined my card. I had not been given an advance to cover any of the out-of-pocket expenses for my temporary duty in England. I used my personal credit card to get out of the airport and on the motorway toward Plymouth. As soon as I pulled out onto the motorway, driving on the left side of the road while struggling to shift the manual transmission and read the road map, I looked in the back seat and saw my daughter's mouth was full of blood. She had managed to yank her first tooth out while sitting in the backseat!

I pulled over to get my daughter situated. I rescued the bloody tooth in her hand before we made our way southwest to Plymouth. I checked into a bed and breakfast until I could sort things out. No one at the agency would take my phone calls. I was on my own and left to stretch my personal expenses. I found a nice two-bedroom flat in Plymouth, not far from the city center in an area of town known as "the Hoe."

St. Andrews Church of England Primary School was literally in my backyard. I immediately enrolled my daughter in school so I wouldn't have to worry about childcare until my course required overnight travel, which was not scheduled until a month later. The experience for my daughter was a significant reason why I wanted this assignment. I not only wanted to invest in my professional development, I also wanted my daughter to adopt the world as her neighborhood. She had been reading since she was three years old. I never felt I was challenged in school, so I was determined to provide her with as many opportunities as possible. I still had faith in the "American Dream." I was happy when I found out that St. Andrews did not use the American grade structure. On the first day of school my daughter was Year 1, Level 1. The next week she was transferred to Year 1, Level 2. By the third week she was in Year 1, Level 3 studying French, Greek mythology, and pre-algebra with eight- and nine-year-olds. It came as no surprise when ten years later she graduated at the age of 16 as valedictorian of her high school class. She set her eyes on Georgetown University very early, and ended up graduating from the School of Foreign Service.

I struggled the entire six months I was in England. Luckily, the exam for the math modules was first. The math modules determined who stayed for the remainder of the program and who would be returning to their countries. I was the only woman. I was the only American. I was the only Black person. And I was the only person who publicly identified with the Christian faith. The rest of the class consisted of naval officers in the Royal Navy, officers from the Netherlands, New Zealand, Australia, and Indonesia. The Long Course extended across the United Kingdom. When I was not in class at HMS Drake in Plymouth, I was at the University of Bristol, the University of Southampton, and the University of Exeter or somewhere in the hills or on the coasts of England getting field experience in hydrographic surveying.

Receiving per diem from NAVOCEANO to accommodate the cost of living in Plymouth was problematic the entire six months. No one from the agency travel office, nor my immediate chain of command, would return my calls. I survived by traveling over five hours north every other weekend to either one of two U.S. Air Force bases, Mildenhall or Lakenheath. Because I was on government orders, I had base access and all the amenities afforded to uniformed personnel. I shopped in the commissary and I was able to access my payroll deposit from the base ATMs. Six months of gloomy-faced Britons and days of overcast and rainy weather did not end soon enough. The financial strain made it difficult to enjoy the experience. The highlights of our time in England were visits to British theaters, visiting the universities of Oxford and Cambridge, a trip to Legoland, and having my parents come to visit.

The Navy's unique contributions to national security stem from the advantages of operating on, under, above and from the sea. This is the message of Forward ... From the Sea. The primary purpose of forward deployed naval forces is to project American power from the sea to influence events ashore in the littoral regions of the world across the operational spectrum of peace, crisis and war. That is what we do. This paper describes how we do it today, and how we will do it in the future.

FORWARD ... FROM THE SEA

THE NAVY OPERATIONAL CONCEPT, MARCH 1997

FORWARD FROM THE SEA

I successfully completed the Royal Navy Hydrographic Long Course at HMS Drake in Plymouth, England; however, the victory was bittersweet. Mentally, physically, and financially I was wrecked. My personal and professional life spiraled out of control. I still beat myself up over whether I should have given up my fight to be accepted into the course after the tremendous effort taken by the agency to prevent my selection. Should I have declined the offer to drag my daughter half-way around the world on half the specified Joint Travel Regulations (JTR) per diem and no dependent allowance? It had taken months for the first payment to be disbursed and every subsequent payment was delinquent. Why had I not returned to the U.S. after I discovered at London's Heathrow Airport that the spending limit on my government-issued Diners Club Card had been reduced? There were several warning signs early on in the assignment that should have prompted me to board a jet and take my hind parts back to the United States.

The IHO Category A status did not garner me a position as a lead hydrographer. In fact, when I returned from England, a travel schedule had already been pre-planned for me and I was scheduled to travel in the same capacity as I had before I even received the GS-12 promotion. A strategic decision had been made while I was away that would keep me out to sea for a minimum of six months a year. I was scheduled for three-months on and three-months off rotations for the next year. As one would expect, my mental and physical health began to suffer from the stress. The day finally came where there was no one available to take my daughter for three months. I knew I had to confront my persecutors, the Commanding Officer, Captain Larry Warrenfeltz, the Executive Officer, Captain Tim McGee, and the senior civilian at the agency and Technical Director, Mr. Bernard Landry, all who collaboratively orchestrated the retributive actions against me. Repeated requests to be removed from the travel schedule long enough to find someone to leave my daughter with for three months were denied. It didn't matter that there were women, white women specifically, who were in the same travel position as me, but who had not traveled in years because they had young children.

One of the Black supervisors in the computer science department was accidently copied on an email correspondence from the Commanding Officer, Captain Warrenfeltz. The email was shared with me under the most covert, swear-to-secret conditions, for fear the supervisor would be implicated and somehow snarled in my affairs with the agency. Captain Warrenfeltz, in his email, where my name appeared on the subject, was very emphatic that the agency would make my life very difficult because of the previous class action grievance I filed. I was "between a rock and a hard place," or as my Momma used to say, "Between the devil and the deep blue sea."

The day I was reprimanded for not fulfilling the duties of my position and refusing to deploy for a three-month tour to the Middle East was one of the darkest days of my life. I had no one to care for my daughter. All I had ever wanted were the ubiquitous opportunities that were promised to me from the time I was a young girl: that if I worked hard and got a good education, I could be and do anything I wanted to be and do. It was the mission that attracted me to the civil service in the first place. I wanted to travel the globe and breach uncharted territories. I wanted to contribute to defending the vital security of our nation. I had invested seven years of education and over ten years traveling in remote areas around the world to areas of ethnic conflict and civil war. I had left my daughter for months at a time, missing her first steps, her first words, and her first tooth.

I had exceeded all that was asked of me. When I was not at sea, I was in the office processing hydrographic data, reading and interpreting side-scan sonar images, plotting acoustic profiles from measurements of conductivity, temperature, and pressure probes. I was taking classes and spending countless hours in the Maury Oceanographic Library delving into the propagation and reflection of sound in water. I carried the mantle of those women who dared to do it all, to be a wife, a mother, and an accomplished professional. Mae Jemison, the Black American astronaut, chemist, and doctor, was my role model. I believed that I didn't have to be just one thing and do the same thing for 30 years and then sit around and wait to die. I could rest when I was dead. I had done everything asked of me, both personally and professionally. In fact, I had done more than most. I had a reputation, which I still carry today, for always doing something to be better and to reach higher goals.

The reprimand, which I received on a Friday, stated that I had

48 hours to report to the ship stationed in Dubai, in the United Arab Emirates. I spent all day Saturday and Sunday cleaning out my desk and loading years of books into milk crates, books I had accumulated during over ten years of government service. I was prepared to be terminated the following week. When Monday arrived, I had every intention to show up for work. I dropped my daughter off at school and headed towards the Stennis Space Center, driving west on Interstate 10, desperately trying to fight the overwhelming sense of anxiety that prevented me from thinking or comprehending. I was so caught up in my thoughts that I passed the Stennis Space Center exit that led to my office. I don't know why, but I just kept driving. I crossed the Pearl River Bridge that separates Mississippi from Louisiana, crossed over the long Pontchartrain Bridge between Slidell and New Orleans, and continued on through Louisiana to Texas. I had enough sense to stop for gas but not enough sense to turn around and go back home or to work. I drove. I cried. I screamed. I felt hopeless. When I realized it was nearing the time to pick my daughter up from school, I was somewhere in Texas. I called my friend Pam and asked her if she would pick up my daughter and take her home with her. I called my frantic parents, poorly explained to them that I was okay, but not okay. I checked into a hotel somewhere in Texas long enough to get a couple of hours sleep. I had cried all I could cry and I needed to get back to my daughter and find out if I had a job.

I drove all the way to Texas and back without ever turning on the radio. I had wrestled with God long enough. My superiors, who thought I was en route to Dubai, were dumbfounded when I reported to work Tuesday morning as if everything was normal. No one said a word to me, not my supervisors, not my coworkers. Even my closest coworkers avoided me, except my friend,

Kathey who had kept my daughter on numerous occasions. She was a single mother as well. Kathey reminded me that God had not blessed me with a child for me to constantly leave her with anybody who could take her. If the toxic work environment were not enough to jolt me, the words Kathey spoke about my daughter hit me like a ton of bricks. I needed a plan, one that allowed me to be the mother God had intended for me to be to a beautiful, precocious little girl and one where I was valued as a human being.

Although suffering mentally and physically and being treated for ulcers and depression, I continued to show up for work. I made copies of all the correspondence with my superiors which documented my employment history, performance evaluations, awards, and the paper trail that led up to the reprimand and conditions of termination. I emailed everything from my work email to my personal email. I requested a copy of my personnel file from Human Resources. I also requested a copy of my medical records from the agency physician who performed our annual physicals to determine if we were fit for sea duty. After 12 years of working for the agency, my records were mysteriously missing.

Eventually, my supervisor was forced to speak to me. He informed me that although I had not been removed from the travel schedule, I had an obscure assignment in a makeshift computer room with two brand new Hewlett Packard (HP) servers. The assignment was to "configure the servers" for processing hydrographic data. I was not provided any information about the servers. I had been the System Manager while deployed on oceanographic vessels at sea, responsible for ingesting, processing, and distributing data from a myriad of oceanographic, bathymetric, and navigation sensors, so I had knowledge of HP administration. However, the bits I knew I had learned from scripts written by more experienced system administrators. The first day in the computer room, I looked around and it was obvious that I had been sentenced to solitary confinement. Were it not for the hum of machinery in the room, the only sounds would have been my deep breaths and grunts trying to understand the requirements for the two brand new HP servers and the actions needed to configure them for processing hydrographic data.

The good news was that I had a job. The not-so-good news was that by giving me a seemingly impossible task, the agency was betting on me hanging myself by failing to complete the assignment. I had successfully completed the Royal Navy Hydrographic Long Course, even when the only other employee sent from the agency had failed to pass the course. I had been a subject matter expert in the littorals for 12 years. As a part of the Total Quality Management team, I had assisted in developing requirements for a new class of oceanographic survey platforms, and I had also provided high-level drawings for those platforms. I was not a computer specialist, but I was equipped with two things. I had taken BASIC, FORTRAN, COBOL, C++ and DOS programming in college, so I knew that somehow, I needed to interface with the operating system of the HP servers. I was also content on staying in God's perfect will. I stumbled on an open brief case containing a set of HP-UX books. It seemed as if someone else had already failed at this assignment. The books kick-started my approach to configuring the servers, enough for me to figure out that I needed HP-UX 11 and the books were written for HP-UX 10. I ordered HP-UX 11 books with my personal funds. The computer room became my personal laboratory and the two HP servers were mine to conquer!

No one came to check on me in the computer room except for the occasional concerned but cautious co-workers, especially my friend Calvin who was like a brother to me. My supervisor never asked for a status. I had bible quotes from my pastor, Robert J. Taylor, in rotation in my head. "You intended to harm me, but God intended it for good!" I was like Daniel in the lions' den. I was like the three Hebrew boys saved from a fiery furnace. I became proficient in HP-UX and in less than a month, I had both servers singing Hallelujah (not literally, of course)!

I juggled the HP server assignment and volunteered for several short-term tours in and around the Gulf of Mexico. I participated in sea trials as well as integration and testing of the ISS-60 integration system which was a server-workstation configuration on a local area network on board the new PATHFINDER class vessels. All the bathymetry, environmental, and navigation data was ingested by ISS-60 and the appropriate algorithms and filters applied that enabled us to plot depth contours of the oceans and littoral zones. Electronic charts were just being developed. So we were the first users of geographic information systems. I had a couple of interesting assignments in the Bahamas and Jamaica. My goal was to get as much sea duty as possible because I was unable to leave my daughter for several months at a time. Thank God for friends, like Denna Richardson, Pamela Clinton Smith, Kathey Dunn, Tabitha Roberts, and Harriette Taylor-Holland, that held me down during those difficult times.

When I was not spending time alone in the server room, and while the agency was still contemplating what to do with me, so sure that the HP assignment would have me write my own termination letter, I was applying for other jobs. Because my experience had been so specialized, it was very difficult to find positions to apply for. I revised my resume several times and decided I would market myself as an Information Technology (IT) specialist in

UNIX administration, marketing the skills I had learned while secluded in the computer room. When I received a call from Allan Darling, offering a job in the Washington, D.C., metro area as a GS-12 IT Specialist, I was afraid he would think I was a fraud. We talked about my experience at sea as a System Manager and the past few months configuring the two HP servers. Before the call ended, Allan offered me the job at the National Weather Service in Silver Spring, Maryland. I had been ostracized for so long, secluded in the computer room, that I had lost confidence in my ability. I had invested all of my human and social capital in NAVOCEANO, the Stennis Space Center, my neighborhood, and all the people I worked with and lived around. Furthermore, I couldn't afford a cross-country move. The experience in England had exhausted my resources. I wanted the job, but the logistics of getting there seemed like an insurmountable challenge.

I asked Allan to give me a couple of days to consider the offer. I hung up the phone and fell to my knees. I asked God to keep me in his perfect will. I was not willing to repeat the fiasco that took place in England. The next day, I approached the Executive Officer, Tim McGee, and asked again for a non-travel billet, at least temporarily so I would not be in jeopardy of losing my job once they decided to let me out of solitary confinement. Captain McGee did not budge an inch and was not amenable to anything I requested. It was then that I focused on my faith. I had shouted and screamed in church about God "making a way out of no way," "opening doors no man can shut," "making my enemies my footstool," and "qualifying those whom he called." Now was the time. I knew it was to time to step out on faith, an unshakable faith, the kind of faith that moves mountains. All the Bible verses I could recall flooded my senses as I dialed Allan's number and accepted the job as an IT

Specialist at the National Weather Service, a subordinate agency of the National Oceanic and Atmospheric Agency (NOAA) in Silver Spring, Maryland. What would happen next was not anticipated and hit me like a whirlwind. I was in full-blown panic mode, not knowing how I would make the transition to D.C. My panic could not match Allan's excitement I detected through the phone.

My phone began to ring nonstop. Someone would be over to pack my things. A truck would come and move everything to Washington, D.C. I would get the locality increase in salary for living in the D.C. metro area. I would also receive per diem, full per diem for six months while I looked for housing. The agency put my house on the market; it would pay all closing costs and would contribute towards the purchase of a new house in the D.C. area. Things happened so fast I didn't have time to think about the magnitude of it all. Ten days after my divorce was final, my daughter and I were headed north to a new life, away from the Gulf of Mexico and the Naval Oceanographic Office; away from the sea.

Allan saw more potential in me than I knew I had. He issued me a $5,000 voucher each year for the two years I worked for him. The vouchers could be applied to any course IBM taught in UNIX administration. Allan would give me a task, like configuring two servers, one as the main FTP server to receive weather data from stations around the world, the other to be a load balancer, sharing the load should the main server reach its full capacity or become inoperable for some reason. Allan grilled me daily on Internet protocols and secure shells for remote access. Allan also gave me a book allowance for which I purchased every O'Reilly book available in UNIX, Internet protocols, Perl scripting, HTML, and C programming. I developed an HTML interface running a Perl script for users who needed to send data to the National Weather

Service. Allan was extremely smart and demanding. He was probably aware of it, so he smoothed the edges by having Friday staff meetings at Starbucks. Most of the time, he gave speeches to boost our morale and encourage us to keep up the good work. After a few months on the job, Allan called me forward during one of the Friday staff meetings and surprised me by presenting me with a certificate and a pin. The Naval Oceanographic Office had submitted my name for the Armed Forces Civilian Service Medal for my contributions to surveys in the Albanian Sea during the Bosnian War!

At the time I received the medal, I was still deeply depressed over what I perceived as a failure in my federal government service as an oceanographer. Honestly, it wasn't until the time of writing this memoir that the significance of the award became apparent to me. In 1998, then President Bill Clinton established the Armed Forces Civilian Service Medal to recognize "*the contributions of the civilian work force in supporting long-term humanitarian or peacekeeping missions. The award is the civilian equivalent of the Armed Forces Service Medal, and the qualification criteria are similar.*"[16] In order to qualify for this medal, civilians must have participated in direct support of a military operation where military members received the Armed Forces Service Medal. Since I have been unable to obtain my civil service records through the Freedom of Information Act (FOIA), as I was told the records were purged, I am grateful to have this physical token of the sacrifice, "heart" work and commitment to the defense of security around the world. This token does, however underscore the importance of carrying into perpetuity the career records and contributions of civil servants who served in combat zones just as is done for the uniformed service members.

I still remember hopelessness in the eyes of two young Albanian girls, no more than 15 or 16 walking the dirt streets of the narrow coastline that borders Albania at the Adriatic Sea. They were just as intrigued by my presence as I was of theirs, in a desolate place that had been ravaged by war. Although the survey crew was there to check on the status of floating acoustic buoys in the Adriatic and to mark their positions, we ventured ashore to witness first-hand the devastation. Before returning to the hydrographic survey launch to complete the mission of positioning the floating instruments which were used to monitor underwater sound, we allowed our senses to experience the aroma of war.

During the two years that I was at the National Weather Service, my younger sister, who had been in the D.C. area several years before I arrived, started and completed a master's in public policy program at George Washington University. I left the National Weather Service to take a GS-13 position at the National Institutes of Health (NIH) as a Solaris SUN administrator in the Laboratory of Brain and Cognition. Motivated by how quickly my sister had received her degree and still regretting that I had never obtained a degree in engineering, I enrolled at Johns Hopkins University and graduated two years later in 2003 with a master's degree in systems engineering. I earned my degree while working full time at NIH, enduring the tragedy of the 9-11 terror attacks, and raising my daughter as a single parent.

September 11 is a pivotal moment in most people's lives. I remember vividly that I was sitting at my desk at NIH, browsing on my computer, when I saw breaking news of the airplanes crashing into the World Trade Center. That is also the day I learned to trust my instincts and rely on my intuition. Even though the nation had not quite figured out what was about to unfold, and even though I was 200 miles away from the events in New York, I hastily shut down computers and packed up to grab my daughter from St.

Bernadette Catholic School in Silver Spring, a mere seven miles from the National Institutes of Health. By the time I got outside, I received a frantic call from my sister, who at the time was working in the U.S. Capitol Building as a speechwriter for Senator Elizabeth Dole. The Capitol was being evacuated because of the threat of a jet headed in that direction to be used as a weapon of mass destruction. I was still in Bethesda, rushing towards my vehicle while talking to my sister on the phone, when I heard what I now know was a hijacked plane being crashed into the Pentagon. It took two hours for me to get to my daughter at her school, even though it was only seven miles away, and I was one of the lucky ones. Many parents were still in gridlocked traffic into the midnight hour trying to pick up their children as the events of that day unfolded.

I finished the master's program at Hopkins four years after leaving NAVOCEANO. My feeling of accomplishment was still not enough to alleviate the pain I felt each time my mind revisited the isolation and condemnation I endured for daring to challenge systemic racism in a government agency. The move to the Washington, D.C., area and my graduate studies at Johns Hopkins allowed me to connect a few dots that had not been clear to me after initially encountering *From the Sea*. In that pivotal document the Navy and Marine Corps envisioned the changes necessary for the naval service to shift priorities away from operations at sea to projecting power and deployment of forces in areas adjacent to the oceans and seas—the littoral regions of the world—where troops and vital security interests were most vulnerable. Four years later I was still determined to adapt professionally as the nation's defense strategy evolved, even if I was not at NAVOCEANO. I started developing strategic goals of my own that aligned with the nation's security and military strategies.

Two years after releasing *From the Sea*, the Naval Service issued an update titled, *Forward, From the Sea* to expand on and translate the vision into objectives for deploying "naval expeditionary forces in peacetime operations, in responding to crises, and in regional conflicts." *Forward* operationalized the global economic, political, and military vital interests of the United States. *Forward* also addressed forward-deployed naval units at sea as the "building blocks" of the peacetime strategy, referring to Aircraft Carrier Battle Groups, surface warships, and ballistic missile defense capabilities.

I convinced myself that my four years in D.C. had been a necessary detour toward reaching my dreams. I had learned more about the defense acquisition process at Hopkins, and I could see my years at NAVOCEANO and my graduate degree complimenting one another. I devised a plan that would enable my past experiences and Hopkins graduate degree in systems engineering to culminate into a yet-to-materialize rewarding career in defense of the nation. Ingalls Shipbuilding, where I had done my cooperative education work while at Mississippi State, had been acquired recently by Northrop Grumman Corporation. I accepted a job with Northrop Grumman as a Systems Engineer. My daughter and I left the D.C. area and returned to my hometown of Moss Point, Mississippi.

My supervisor at Northrop, Bob Rifley ("Rif"), a guitar playing rocket scientist, put me to work right away on the U.S. Navy's DD(X) program. I was familiar with the design of the surface combatant, referred to as DD-21 from a class exercise in my graduate studies. DD(X) was a concept that emerged from the maritime strategy initiated by *From the Sea* and operationalized in *Forward* to shift naval priorities and equip maritime forces in the littorals.

Ingalls Shipbuilding had been "building great ships" for over 50 years, and thanks to Trent Lott, provided a stable revenue stream for Mississippi. The Navy typically provided Ingalls with a mature design and Ingalls maintained the craftspeople and expertise to keep the Navy's fleet well equipped for war at sea. The award of the DD(X) design posed a challenge to Northrop's newly acquired shipyard, which was an opportunity for me. Defense acquisition regulations mandated rigorous and repeatable systems engineering processes be implemented on the DD(X) design. The design also included several innovative approaches to shipbuilding in its hull form, power systems, and missile launches. I hit the ground running with Rif pushing me along the way. I drafted many of the plans and processes for DD(X) and collaborated with industry peers from Raytheon to develop concept designs and use cases for the ship's integrated power system. The DD(X) design was intended to be the standard for "Total Ship Systems," modular functionality, open architectures, and integrated computing. I traveled extensively between Mississippi and Raytheon's facilities in New Jersey, occasionally flying to other subcontract facilities where smaller tasks were being performed.

For the first year, I was engrossed in my work and enjoyed having most nights at home with my daughter and weekends eating at my Momma's kitchen table. Northrop was awarded another contract to modernize the Coast Guard's aging fleet of vessels. Rif had enough confidence in me after only a year at Northrop to promote me to Engineering Supervisor and assign me as co-lead with Greg Carithers to the Coast Guard's Deepwater program. I'm not going to lie. I thought I was living the dream. My day would start by getting my daughter off to school and then driving to the Trent Lott International Airport a short distance from my house.

I would board a turboprop aircraft along with Greg and four other coworkers to commute to New Orleans, where the 400 foot Coast Guard cutters were being designed. Greg was a pure computer scientist, having been laid off from Scott Paper Company prior to joining Northrop Grumman Ship Systems. Rif, who was an extremely good manager of both people and technology, was wise to pair me and Greg. I leveraged Greg's expertise while offering a maritime background and a systems engineering degree that he did not have. Greg and I had a lot in common. He was focused, somewhat of an introvert, and a hopeless workaholic.

Rif's confidence in the team and the commuter plane rides to work every day blinded me. Greg was shrewder than I had given him credit for. In fact, Greg turned out to be conniving, passive-aggressive, and manipulative. He was a master at Microsoft Excel and I played "follow the leader" trying to figure out how data in spreadsheets I created would mysteriously change. Each time Greg had to come along and "save the day" by correcting my work. I would spend hours burning the midnight oil checking and rechecking my work. Greg prepared all our presentations and I would stand to present and my data would not reflect the reality of my analysis. Reverting back to my days at NAVOCEANO when I taught myself how to administer the HP servers, I ordered Microsoft Excel books. That's how I discovered that Greg had created macros in the EXCEL spreadsheets with Visual Basic programming. My ears still turn into flames when I think about the lengths that people like Greg will go to in order to sabotage another coworker. This tactic, I would learn, is commonly used to devalue others, especially people like me who threaten their perceptions of superiority and devaluation of "the other." Because of Greg, the smell of superiority has been seared into my consciousness. I recognize its aroma in the eyes, attitudes, and behaviors of persons when I walk into a room. I can instantly distinguish the stench of those who see me as an undervalued threat, as if I don't belong. This scenario would play out time and time again for the rest of my career.

Rather than challenge Greg directly, I took a less aggressive approach. Determined not to fight, I dove into Excel how-to books and learned how to create my own macros. I locked down and protected my spreadsheets so that I would know when macros were running in the background. I also learned how to write my own scripts and maintain control of my data. I still find it amusing that Greg never mentioned his inability to modify my spreadsheets after I blocked him. I guess that's because he advanced his tactics and went to the next-level scheme, such as sending emails at three in the morning and setting up meeting invites for 7:00 a.m. the same day. He would also conveniently forget to add my name to distribution lists.

Eventually, my choleric temperament got the best of me and I had to convene a "Come to Jesus" meeting with Greg and Rif. Rif agreed that the late night and weekend work assignments were problematic, but by then, I had lost all trust and respect for Greg. His sinister behavior placed unnecessary tension on our working relationship. Rif sensed the tension and delineated the lines in our responsibility. Greg would lead the National Security Cutter (NSC) effort and I would lead the Fast Response Cutter (FRC) program. This new scope in responsibility expanded the number of team members and products I was accountable for, which included a contingent of employees working in the Washington, D.C. metro area and then required that I travel frequently to D.C. When I left home after waking up my daughter, I would drive to Trent Lott International Airport to board a larger, more luxurious commuter plane and fly from Mississippi to D.C. on Northrop's corporate fleet of aircraft and then be back home before the sun went down.

The Coast Guard's Deepwater program was an unusual amalgamation of government acquisition contractors. Northrop Grumman entered into a joint venture with Lockheed Martin to form Integrated Coast Guard Systems (ICGS). ICGS was awarded a government contract to design and build a fleet of cutters to replace the Coast Guard's aging fleet, among those were the NSC, FRC, and the Offshore Patrol Cutter (OPC). ICGS then subcontracted tasks to other large and small companies. The headquarters for ICGS was located in Arlington, Virginia, and was collocated with Coast Guard counterparts. ICGS was several years into the joint venture without reconciling how the two prime contractors, Northrop and Lockheed, would integrate their disparate processes. Changing culture doesn't happen overnight. My Daddy's repetitive mantra about not being able to change people, but to change your approach applied to corporate American culture as well. But the Coast Guard Deepwater program was a risky venture from the start. The Coast Guard needed an integrator to manage the people, systems, and processes necessary to mitigate risks and build platforms to replace the Coast Guard's aging fleet. The acquisition strategy was innovative, but it was also unrealistic and risk-prone from its inception. The multibillion dollar contract awarded to a hastily formed joint venture between two competing contractors, Northrop Grumman and Lockheed Martin, proved to be a disaster. Findings by the Government Accounting Office (GAO) in 2004 stated:

Over a year and a half into the Deepwater contract, the key components needed to manage the program and oversee the system integrator's performance have not been effectively implemented. Integrated product teams, the Coast Guard's primary tool for overseeing the system integrator, have struggled to effectively collaborate and accomplish their missions. They have been hampered by changing membership, understaffing, insufficient training, and inadequate communication among members. In addition, the Coast Guard has not adequately addressed the frequent turnover of personnel in the program and the transition from existing to Deepwater assets.[17]

Even though ICGS processes were drafted and approved by ICGS leadership made up of both Northrop and Lockheed management, Lockheed continued marching forward with processes it had implemented and proved successful on previous contracts for integrating Command, Control, Computers, Communications, Intelligence, Surveillance, and Reconnaissance (C4ISR) systems. On the other hand, Northrop pressed forward with its Total Ship Systems approach to integrating hull, mechanical, and electrical (HME) systems and building great ships. My commute between D.C. and Mississippi would consist of capturing the different perspectives I had encountered between Lockheed's engineers and Northrop's engineers. Unlike my Northrop counterparts, I was actually willing to listen to Lockheed's engineers and try to understand why they opposed our Total Ship Systems architecture and functional allocations. The fundamental issue and source of the conflict between the two contracting giants all boiled down to the absence of an ICGS taxonomy. Both sides were using the same words but referring to different concepts. This basic discrepancy resulted in almost intractable conflict between the two sides, especially when we had to negotiate with the Coast Guard contracting office about how to share the $17 billion dollar purse.

Establishing an ICGS taxonomy for designing the Coast Guard cutters was just the beginning. It was important for me to foster a working relationship with Lockheed, one that did not compromise my allegiance to Northrop or my supervisor. I learned that the bottom line for management in business is future dollars. On paper, ICGS was a joint venture. In reality, the two companies sparred for position on future work almost daily. Every decision made was a result of intense internal deliberations of "if" Northrop does this, "then" Lockheed would do that. My Lockheed counterpart and I found common ground on identifying the systemic failures of the C4ISR and HME integration issues, but he had the same challenges as I did in trying to convince Lockheed management to rethink their methods, even if for the good of the program. Both sides saw compromise in terms of loss future dollars.

Once again, I turned to books and the classroom. I engaged in a self-paced crash course in enterprise design and architecture. I understood ship system architecture and its taxonomies. My Lockheed counterpart helped me to transition my knowledge of ship systems into the C4ISR domain, and the books I read on transforming enterprises allowed me to devise and implement an approach that truly integrated the domains of the two joint venture partners. I represented Northrop at the ICGS System Architecture working groups and assisted in the development of many architecture artifacts for the design of the Coast Guard cutters. I spent so much time in D.C. that Northrop eventually relocated me to the area and I was able to become the Lead System Architect with all the duties and responsibilities of the Deputy Technical Director. However, I was never officially given the title of Deputy Technical Director. I managed a geographically dispersed team of engineers and I had overall responsibility for systems engineering and architecture development, safety, risk, quality, and test engineering.

The day it was announced that another gentlemen would be assuming the role as Deputy Technical Director, I grasped the intensity with which I was undervalued. Despite the fact that I was promised the job and doing the job quite successfully, I was expendable. This was the first, but not the last, time I would learn that there are two options when government contractors are faced with reductions in funding. Employees are either laid off or transferred to other programs. But that was not the real revelation. The real revelation was the structural hierarchy and valuation for reassignment versus termination. The informal, yet institutionalized rule that determines who gets laid off and who gets transferred to other programs usually begins by transferring white male veterans. After that, any remaining nonveteran white males are transferred, which is then followed by the "FBI" method. "Friends, Brothers, and In-laws" are given secure positions on other programs, even if it means sacrificing other high-performing employees like me, including white women. Rarely are any positions available by the time the "FBI" has taken all that remains on other programs. Companies like Northrop maintain a list of employees whose programs have lost funding and the employees who are looking for programs to transfer to. This gives the government the illusion that the company takes care of its employees and that there is equity when reductions in funding lead to layoff. Smoke and mirrors.

The Deputy Technical Director rarely came to work at the ICGS offices on Wilson Boulevard in Arlington. Although he had the title, I still had a nice window office on the top floor and I was still evangelizing for a more integrated approach to C4ISR and HME design of the Coast Guard cutters. When the Deputy

Technical Director did show up for work, he rarely had anything to say, at least not about the program unless it was to ask someone what had taken place in his absence. Although I was disappointed that he held the title for the job I was doing, I was content that I was doing meaningful work that I enjoyed. I had a healthy work-life balance. I lived a short bicycle ride from work. I continued trying to facilitate better communication between the industry partners by developing the taxonomy and enterprise architecture framework, leveraging the Department of Defense Architecture Framework (DODAF). In 2006, I presented a paper to the American Society of Naval Engineers' "Engineering the Total Ship (ETS) Symposium" on the challenges that are fundamental to the ICGS enterprise. I was also a presenter at the System of Systems Community of Interest Symposium at Fort Belvoir that same year. I made a good case for how to mitigate the enterprise issues and avoid the inevitable, the demise of ICGS.

After many technical deficiencies, cost overruns, and schedule slips the government eventually pulled the Deepwater program away from ICGS control. However, I managed to make an escape before being among the Lockheed and Northrop employees who were either transferred or terminated, depending on where they fell in the valuation hierarchy. Since there was zero probability that I could ever be a white male, I set my sights on owning my own company, where I could dedicate long hours and brain power to my own business.

The United States Navy, Marine Corps, and Coast Guard are our Nation's first line of defense, often far from our shores. As such, maintaining America's leadership role in the world requires our Nation's Sea Services to return to our maritime strategy on occasion and reassess our approach to shifting relationships and global responsibilities. This necessary review has affirmed our focus on providing presence around the world in order to ensure stability, build on our relationships with allies and partners, prevent wars, and provide our Nation's leaders with options in times of crisis. It has confirmed our continued commitment to maintain the combat power necessary to deter potential adversaries and to fight and win when required.

A COOPERATIVE STRATEGY FOR 21ST CENTURY SEAPOWER

SECRETARY OF THE NAVY RAY MABUS, MARCH 2015

AN (UN) COOPERATIVE STRATEGY FOR THE 21ST CENTURY

Among Daddy's many sayings was, "Get all you can, and can all you get." Another one of Daddy's sayings was, "Once you get it in your head, they can't take it from you." Daddy's philosophies were always swirling around in my head. Subconsciously, I was compelled to read everything I could about my profession, the institutions that governed it, and the sociopolitical aspects of why things were the way they were. I was on a constant search for the cause and effect of things culturally and politically in the institutions that guided the development of U.S. warfare technologies and their deployment. There had to be, in my mind, a point of increasing returns on the investments I made into myself and the activities I was exposed to. I had matriculated in British universities, received extensive government training, graduated from the prestigious Johns Hopkins University Whiting School of Engineering. I had a unique insider perspective on the science and technology, the cultural inequities, and the bureaucracy of federal government machinations, yet on the ground, in the trenches, I was still forced to face the same battles of systemic racism and devaluation.

By the time the Navy, Marines, and Coast Guard released *A Cooperative Strategy for the 21st Century* in 2007, the first maritime strategy to be signed by all three leaders of the U.S. maritime services, I had baptized myself into the naval culture and knew well in advance of that document's release and what to expect in it. Despite being often overlooked, forced to stand against the wall, and soak in all the knowledge emanating from the room, I joined the American Society of Naval Engineers (ASNE), The Society of Naval Architects and Marine Engineers (SNAME), The International Council on Systems Engineering (INCOSE), the Navy League, and I volunteered to serve on committees for the National Defense Industrial Association (NDIA). NDIA's website describes the organization as "*a non-partisan, non-profit, educational association that has been designated by the IRS as a 501(c)3 nonprofit organization* - not a lobby firm - *and was founded to educate its constituencies on all aspects of national security.*"[18] It would have served the organization better not to use bold type and try to disassociate as a lobbying firm. NDIA's methods and techniques are lobbyist-like. As a nongovernmental organization in the Washington metro, the membership is represented by businesses whose main interest is national security. The lobbyist-like organization hosts conferences and symposiums to allow government and defense industry leaders to make presentations, network, and give the informed contractor a leading edge on upcoming acquisition strategies and priorities. In reciprocal spirit, businesses preposition themselves for future contracts with the government. NDIA's website describes its policy-making activities. *"The NDIA Policy team monitors, advocates for, and educates government stakeholders on, policy matters of importance to the defense industrial base. Our mission is to ensure the continued existence of a viable, competitive national technology and industrial base, strengthen the government-industry partnership through dialogue, and provide interaction between the legislative, executive, and judicial branches."* Similar to lobbyists, NDIA facilitates the building of alliances and serves as a champion to influence decision-makers. NDIA's conferences and symposiums are offered at a hefty registration fee, and it should be noted that the government decision-makers who present on their panels are not allowed to accept an honorarium. NDIA registration fees for events range from $500 to $2,000, depending on whether you represent academia, government, or industry and how early you register.

I also attended the annual Sea Air Space (SAS) Exposition hosted by the Navy League. Although SAS is popular among government employees and contractors for the massive defense and maritime exhibits and free trinkets like stress balls, ink pens, and conference bags, I attended SAS every year to listen to the panelists and to ear hustle on the happy hour discussions to find out the naval services' important strategies and policies.

In the unequal, uncooperative federal government sandbox, if you wait until a strategy is released, you are already four or five years behind the competition. The year before the *Cooperative Strategy* was released, I presented my business plan to the Small Business Association (SBA) and embarked on a self-paced study of doing business with the federal government. I had experience at Northrop preparing basis of estimates, drafting statements of work, developing requirements, managing subcontracts, and negotiating with the government contracting officers.

I developed a strategic vision for my business that closely aligned with the impending *Cooperative Strategy*. I used all the buzz words like *joint, integrated, securing the homeland and citizens, interdependent, interoperable,* and more importantly, I emphasized how my company would add value to maritime modernization and improved integration and interoperability, which were foundational to the core capabilities described in the strategy. I drafted an original plan titled, "Navy Open Architecture Systems Engineering and Requirements Management Plan (NOA SERMP)," which collated and synthesized all the lessons I learned from the risks and failures I observed in the modular and open architecture design efforts of the Pathfinder class of oceanographic vessels, the DD(X) surface combatant, the Coast Guard cutters, and my engagement with various Department of Defense Architecture Framework and Navy Open Architecture working groups. At that time, there was not a similar comprehensive approach that could be tailored for the type of cooperative efforts the strategy required and that would be compatible with the DoD-burdened acquisition process for deploying technologies to the warfighter. The ICGS and Deepwater program had unveiled a lot of unknown risks, but the joint venture was dismantled before any of the conceptual processes and methods I developed had been officially adopted.

By virtue of my involvement in so many professional organizations, I had amassed a virtual rolodex of who's who and what's what in the defense and Navy upper echelons. I maintained information on the government agencies I needed to market my services to. I also maintained files on my competitors, particularly on and near the Gulf Coast of Mississippi, since that is where I planned to base the company even though I envisioned having nationwide government clients. I found only one competitor, a small business, targeting the Navy and the shipbuilding industry, collocated on the Mississippi Gulf Coast and within the nation's capital. Gryphon Technologies marketed itself as a small business and checked off several of the government set-aside requirements of being either veteran-owned, minority-owned, or woman-owned. I conducted routine surveillance "drive-bys" of Gryphon's Gulf Coast office. The more I researched the competition, the more it seemed like Gryphon was a large business dressed in a small business outfit. The SBA determines business size based on annual receipts and the number of employees. My research revealed that Gryphon exceeded both those limits, but that revelation also led to the discovery that they were in the SBA's 8(a) program, which is designed to help socially and economically disadvantaged businesses by setting aside certain contracts for qualified businesses.[19] I added the 8(a) program to my strategic plan, a plan I updated every two or three months as a way to measure progress toward my stated goals. The good news was that although Gryphon was the company I needed to compete against, I was not really in their league because they would be soon graduating from the 8(a) program.

I had not yet quit my job. I was delaying that for when I actually received a government contract. I burned the midnight oil after work, crafting lengthy responses to government requests for proposals. I focused on a way to market myself before announcing my business. I had connections through the various organizations I supported. I needed a way to introduce the collaborative ideas I had for meeting critical defense and homeland security needs.

The mistakes of Hurricane Katrina were still a contentious subject, and it was an ideal cause to construct the idea for a Gulf Coast Community of Interest. I conceived the idea for a regional symposium on the Gulf Coast. I contacted people I had worked with on various committees for NDIA and was grateful for the invite to their Arlington office in order to pitch the idea of a symposium cosponsored with NDIA. Although I received a lot of head nods and smiles, NDIA declined to cosponsor, but then sponsored a very similar symposium with the concepts I proposed.

Despite NDIA's rejection, I pressed forward since I was targeting the Gulf Coast corridor. I felt confident that I stood a better chance of forging relationships on the coast than with any organization in and around the beltway. I invited decision-makers from federal, state, and local governments who had a stake in sharing disaster preparedness and response information. I developed the framework for the Gulf Coast Community of Interest. I established a nonprofit entity, Architecture Systems & Technologies. The Gulf Coast Integration, Information and Interoperability Symposium was held on February 21, 2007, at Pelican Landing Conference Center in Moss Point, Mississippi. The Office of the Deputy Chief Information Officer, Assistant Secretary of Defense for Networks & Information Integration (ASD(NII)), was in the beginning stages of an information sharing pilot program and provided valuable information to be distributed to the attendees for enabling such sharing through communities of interest. Philip Teel, then President of Northrop Grumman Ship Systems, was the keynote speaker. Then U.S. Navy Captain Jim Shannon provided a brief on the Navy's open architecture initiatives. Chris Gunderson, a faculty member of the Naval Postgraduate School, John Weathersby from the Open Source Software Institute, and U.S.

Navy Captain Mary Dexter, Commanding Officer of SUPSHIP Pascagoula, were among the panelists for the symposium. The symposium was well attended by engineers, first responders, state and local agencies, and other interested individuals from across the I-10 corridor from Louisiana to Florida; the groundwork was laid for establishing the community of interest. We held weekly teleconferences, but interest began to wane as state and federal agencies were moving much faster than I to establish an infrastructure for sharing information.

The demise of the community of interest was no deterrence. The purpose of the symposium was to market my abilities and get exposure for my business ideas, while not posing a conflict of interest for my job at Northrop. I started the business under the auspices of a nonprofit with the full intention to reorganize as a for-profit business at a later date. I had to be very careful that I didn't present a conflict of interest with my employer. I had lofty aspirations. I believed in my abilities. I was innovative and daring enough to believe that I had what it took to launch a successful government contracting firm. I had witnessed others do the same with less experience, less resources, less education, and fewer contacts. I studied the successful businesses. I knew Northrop Grumman's business development strategies. I had worked with small business subcontractors and read through their marketing, technical, and financial approaches. I closed out the affairs of the community of interest with most of the participants expressing interests in working with me on future endeavors should I pursue a business with the concepts and frameworks I proposed. I accomplished my objective and I was able to include those efforts in the response to a proposal the Navy issued for small businesses to provide services under the umbrella of the Seaport-e contracting vehicle.

Seaport-e was the acronym for the electronic platform and contracting vehicle the U.S. Navy established to enable small businesses to compete for contracts without having to compete against larger, more resourceful contractors for engineering, financial management, and program management tasks.[20] Securing a Seaport-e contract with the Navy is equivalent to a having an invitation to the dance, but you can't just show up to the dance. You have to prove to the Navy that you are worthy of the invitation, then they send you a ticket. Seaport-e Requests for Proposals (RFP) only occur once or twice per year. The required response is lengthy, detailed, cumbersome, and requires the assistance of someone with expert government proposal writing skills as well as someone who understands government cost accounting, the Federal Acquisition Regulations (FAR), and the Defense Accounting and Audit Agency (DCAA) requirements.

When I think back on those days, I'm not sure how I survived, but when I look back over my life, I have always extended myself far beyond the average person in order to be successful. I wanted to be successful more than I wanted to breathe. I still believed in the American dream that my hard work would pay off. I was a single parent. I was managing a geographically dispersed team of engineers, traveling from coast to coast, and somehow, I managed to jump head first into learning and applying skills I had acquired with Northrop and successfully submitted a proposal to do business with the United States Navy.

"This email is to notify you that your company has received a prime contract award as a result of the Seaport e Rolling Admissions solicitation."

I remember the day the email showed up in my inbox. The subject was "Seaport e Rolling Admissions Award Announcement." It was 4:19 p.m., May 23, 2007. I was sitting on the sofa in the house I rented in Arlington after returning to the D.C. area to be the Lead System Architect for the Deepwater program. The earth stood still for a few seconds. I could feel my heart beating hard and fast and loud as if it would pop out of my chest at any moment. Did this mean what I thought it meant? Had I actually single-handedly submitted a proposal and been awarded a Seaport-e indefinite delivery indefinite quantity (IDIQ) contract award for a couple of million dollars? I remember calling my Momma and screaming into the phone. She couldn't understand one word of what I was saying. I could hear my Daddy yelling in the background, "Tell me! Tell me!"

I rode that high for days, because there was a lot to read and comprehend in the Seaport-e contract and even more for me to learn about doing business with the Navy. As a Seaport-e contract holder, I received access to a database of past, present, and upcoming contracting tasks. The challenge then was to identify tasks that I had the expertise to perform, as well as to identify other Seaport-e contract holders to team with and then submit bids to win the task order! Few Seaport-e task orders were written for a single contractor. Teaming is an integral part of the Seaport-e vehicle. For the next several months I amassed several teaming partners from both large and small businesses, still not announcing my resignation from Northrop. Needless to say, this became problematic. Although I had not received any money from the Navy, I recognized that my activities were skirting too close to Northrop's bailiwick.

A couple of months before the *Cooperative Strategy* was to be released, I had enough teaming partners to start bidding on

Seaport-e task orders. I announced to Northrop my plans to start my own business. I left on good terms with Rif and the rest of my superiors and bid farewell to Northrop with the possibility of doing business with the company when the terms of my non-compete agreement had been fulfilled.

At the time, it seemed like I made all the right decisions. When my daughter was six she had attended a computer camp at Georgetown University. She pledged to attend Georgetown, even at that young age. I promised her if she could get in, I would make it happen. She submitted an application for early decision in the fall of her senior year and it was put up or shut up time for me. If Georgetown thought she was good enough to get in, there was no way I was going to renege on my promise. Even though lack of finances was not an issue, her acceptance came during the first time since I was twelve years old that I did not have a steady income. A business owner can prepare fiscally for negative cash flow, but there is no way to mitigate the emotional risks. Getting a Seaport-e task order proved to be more difficult than I anticipated.

Business development consumed 24 hours a day, 7 day a week. Every person I met and everywhere I went, I was focused on making connections with people and institutions to advance my business goals and position me to win government contracts. The first revenue generating contract I acquired was as a subcontractor to the Department of Homeland Security (DHS). I noticed there were similarities in how I acquired the DHS subcontract with how I transitioned from NAVOCEANO to the National Weather Service. It was the ancillary skills of administering servers that opened doors for me when I transferred from NAVOCEANO to the National Weather Service, not necessarily my 12 years of experience as an oceanographer. I was subcontracted to be the

Program Manager for DHS Information Sharing and Collaboration Council (ISCC) in the Office of Intelligence and Analyst. I was able to get that work as a result of my efforts establishing the Gulf Coast Community of Interest. Keeping a list of my skills separate from my resume, a list that could quickly be updated regularly as I took on new tasks, became a valuable tool for acquiring new work. When I submitted a résumé I tailored the resume for the job by including all the skills that seemed appropriate for the job vacancy. This adjustment allowed me to market myself as a business. I was no longer someone looking for a job. My previous jobs were considered past performance. I marketed my skills the way I marketed my business.

A resume highlighting my job titles and employers would not have afforded me an insider view of the cultures within several different government spaces, particularly the Department of Defense and the newly formed Department of Homeland Security. As the DHS ISCC Program Manager, I facilitated the development of the information sharing framework and enterprise architecture for 22 DHS agencies. I was surprised that the Department of Homeland Security was just as disconnected as it had been a few years earlier when I was on the Deepwater program. DHS, by mandate of former President George W. Bush, had placed 22 previously disparate intelligence and law enforcement agencies under one authority to improve information sharing among the federal, state, local, and tribal agencies. Poor information sharing was to blame for the failure to foil the 9/11 terrorists plot and for the poor response after Hurricane Katrina. I assisted in developing a taxonomy for the DHS agencies, an architecture for the Homeland Security Information Network (HSIN), and performance measurements to assess the success of DHS information sharing.

Later, I transitioned to the Joint Non-Lethal Weapons (JNLW) Agency at Quantico Marine Corps Base, where I facilitated joint activities of all the military services for the Active Denial System (ADS), a directed energy nonlethal weapons research technology. JNLW was tasked to mature the ADS technology and enable it to be deployed in the Global War on Terror (GWOT). I liaised with Combatant Commands and Pentagon officials to overcome the technical, programmatic, and financial risks of directed energy technology. I became a liaison for the U.S. Navy at the Missile Defense Agency (MDA) where I oversaw the activities of the Small Business Innovative Research (SBIR) program, focused on interceptor technologies for ballistic missiles. I also developed a technology roadmap for inserting new technologies into the ballistic missile defense program. General Dynamics subcontracted me to develop architecture, requirements, conceptual design, and a high-level implementation plan to integrate the Coast Guard's Maritime Security Risk Assessment Model (MSRAM) into Coast Guard command center applications in support of tactical decision-making.

I managed to get my daughter through Georgetown without her having to take out those dreaded student loans, at least for the first three years. However, getting a task order award from the Navy through the Seaport-e contracting vehicle proved to be near impossible. I submitted an untold number of proposals as a prime contractor and I was a subcontractor on several proposal teams, but none of those efforts materialized into any portion of the $2 million designated by the Seaport-e contract.

I had one Hail Mary before giving up on Seaport-e. I was teamed with Alion Science and Technology for a Seaport-e task order. I worked with Alion on developing the proposal. My company's past performance, the Navy Open Architecture Systems

Engineering and Requirements Management Plan that I developed, and a portion of the level of effort my company would provide was included in Alion's technical and cost approach submitted to the government. Alion's contracting office sent a subcontract agreement for me to sign, which designated that my company would have six employees allocated for the contract if Alion was awarded the task order. I survived another one of DCAA's tedious audits of my business financial system to make sure I could adequately manage cost accounting methods.

I learned only by checking the Seaport-e portal that Alion had won the task order on which I teamed with them and assisted in the proposal development. I never received a call from Alion, not an email, text, nor postal service mail. My calls to Alion went unanswered. I had contingent hire agreements from potential employees. I had a lease pending for office space. When Alion decided to respond, they informed me that the subcontract agreement "guaranteeing" my company funding for six employees had never been signed by them, intentionally! The agreement had not been executed. I had nothing left in me to fight Alion, neither energy nor financial resources. I asked the Navy's Seaport-e ombudsman if I had a recourse. There was an appeal process, of course, but given my company's limited past performance as a government contractor, to appeal would surely blackball me from future government contracts.

Seaport-e and Alion, be damned!

Then, there was my Daddy's aphorism again, ringing in my ear, "The road to hell is paved with good intentions." I had translated my vision into a strategic plan. I had executed the plan with $50,000 of my own resources, an untold number of sleepless nights, and a ton of sweat equity. So, committed to my strategic plan, I made good on an earlier goal to apply for the SBA's 8(a) program. After expending resources to learn how the program works and how to submit a credible application, the SBA determined that I didn't have sufficient revenue to qualify. If I had sufficient revenue, I wouldn't need the SBA program! Despite the fact that I am a hopeless introvert who just happens to have extrovert skills, I traveled across the country to endure numerous networking sessions with all the professional organizations I had joined and the many government information sessions I attended where the contracting officers went through the motions to describe the government's acquisition goals and procurement vehicles. I shared my credentials and insights with established businesses that needed to show they had minority businesses as teaming partners. Yet, I had to accept the reality that I was at a crossroads. I didn't want to think of it as a dead-end because I still believed, at that time, that I had choices, although my disappointment and frustration were leading me to a conclusion that the American dream was definitely not equitable. There are an untold number of people who have the ability and the determination to be all that they can be, yet they are driven into a corner of discouragement and despair (not to mention financial ruin) by institutional and systemic roadblocks. My education, experiences, and dreams had taken me about as far as I would go, but I had not accepted it then. The ranks of power and the circle of privilege were not easily broken.

Yet, my own internal fortitude kept telling me that this was not the end of the road for me. I still had an ounce of hope that I would land on all fours in a place suitable for the tremendous amount of experience I had acquired. By re-reading the *Cooperative Strategy* and reordering my priorities, I still had confidence in my marketable skills. I wrestled with how to once again channel my anger and failures into a good news story. I had absolute rage at yet another impasse with unethical behavior, being chewed and spit out by an established norm in the circles in which I traveled, circles that were unscrupulous and definitely not cooperative, in my opinion. I had enough evidence to make a case that dodgy contracting methods existed to usurp federal regulations, which required large businesses to set aside a certain percentage of their contracting dollars. The status quo was to amass a team of woman-owned, minority-owned, veteran-owned, HUB zone,[21] and disabled veteran-owned businesses to develop a response to government proposals, which on the surface satisfied all the requirements levied on large businesses to make opportunities available to small, disadvantaged businesses.

What happens in reality is that large businesses establish relationships with a very, very small number of small businesses for which they have a reciprocal relationship of "you scratch my back and I'll scratch yours." If the small business wins a task, the small business gets 51 percent of the work and the large business that propped them up receives 49 percent. If the large business wins a task, the large business takes 51 percent and spreads the remaining 49 percent of contract dollars over the small subset of contractors that would scratch their backs in return. Very seldom, if ever, is this circle broken.

The other tactic large businesses use is to capitalize on the fact that small businesses are so desperate for contracting dollars that they will basically work for free during the proposal process. While large businesses can free their employees to perform on actual funded contracts, small aspiring contractors spend countless hours drafting proposals, searching for qualified contingent hires (at a cost), and doing the market intel necessary to position the large business for a win. Some businesses will even go so far as to allow small businesses to host proposal status reviews at their facilities, giving the small business the illusion that a contract win assures them new business. In most cases, this never materializes, at least not at the magnitude expected. Tossing $100,000 to a struggling business is peanuts to a business having landed a multimillion dollar contract, and barely enough to compensate the small business for the resources expended to shore up the win.

The good news was that my daughter graduated from Georgetown. The not-so-good news was that I was learning just how uncooperative federal government contracting could be. I don't attribute the series of unfortunate events that I experienced to the competitive nature of doing business with the government. For all the regulations and contracting rules put in place by the federal government, there is very little recourse to hold contractors accountable when the rules are broken. I have witnessed one too many small businesses nearly or completely go underwater because of questionable contracting practices by federal government contracting giants like Alion Science and Technology and the nebulous family of Alaskan Native Corporations (ANCs). SBA created special rights for Alaskan Native Corporations on the premise that Indian reservations and Alaskan Native Villages with high unemployment and poverty levels have the privilege of using the *"...Federal government's massive procurement activity to help jump-start reservation economies, Congress has given tribes and Alaska Native Corporations (ANCs) unique rights in the Federal procurement process."*[22] In August 2009, Washington Technology ranked the largest ANCs and tribally-owned corporations in the government information technology market. Of the nearly $3 billion generated by these top ANCs, an average of 82% of that revenue came from set-aside contracts, meaning little to no competition and in most cases sole sourced (that is, handed to them).[23] The basis for preferential treatment of ANCs troubles me, since the 2016 Census statistics show that the percentage of people in poverty in Alaska was at 11%, while the percentage of people in poverty in my home state of Mississippi, notoriously last on the poverty rankings, exceeded 20%.[24]

I could have allowed the pendulum to continue swinging, chasing after Seaport-e tasks, but instead I decided to slow the pendulum balls down and get a stable job, one with a W2 and not a 1099 or business-to-business arrangement. I still had a few business partners for whom I did proposal writing and that provided a decent amount of income to sustain me while I looked for a job. Maybe I was stuck on stupid, believing the hype of the C*ooperative Strategy* would equate to opportunities for my background, but I set my sights on the Pentagon, the headquarters for the Department of Defense. I had covered the lifecycle of defense design and integration, from identifying mission needs to translating those needs into requirements, allocating functions to hardware and software systems, to the actual fielding of innovative technologies. I knew defense, Navy, and joint cultures and personalities as well as the relationships between the services. Engineering innovative technologies was my sweet spot. There was something about having to be creative and methodical at the same time that excited me. It made a boring job kind of exciting. A self-professed geek,

I liked applying structure and rigor to designing and integrating technologies that had tangible benefits to the warfighter and contributed to defending the nation's security interests. Everything I had done up until that point had prepared me to be able to execute the dollars and tasks developed after of years of planning, programming, and budgeting orchestrated by talking heads in the Pentagon. I wanted to be a talking head. I wanted a piece of the pie that developed strategy and policy. I wanted to be able to follow the money allocated to the warfighter, opportunities that had eluded me when I was chasing Seaport-e crumbs.

The Department of Defense recognizes the need to explore and adopt options that will preserve U.S. ability to project power and maintain freedom of action in the global commons. In July 2009, the Secretary of Defense directed the Departments of the Navy and the Air Force to address this challenge and to embark on a new operational concept called Air-Sea Battle (ASB). Since then, the U.S. Army, Marine Corps, Navy, and Air Force have collaborated in new and innovative ways to address the anti-access/area denial (A2/AD) military problem set.

AIR-SEA BATTLE -

SERVICE COLLABORATION TO ADDRESS ANTI-ACCESS & AREA DENIAL CHALLENGES.

Excerpt from an unclassified summary of the classified Air-Sea Battle Concept, version 9.0, dated May 12 and the Air-Sea Battle Master Implementation Plan (FY13), dated Sept. 12.

ANTI-ACCESS/ AREA DENIAL

Ironically, while I was experiencing the divide between my dreams and my own post-racial reality, my childhood neighbor and Professor of Religion and African-American Studies at Princeton University, Eddie Glaude, was writing about how race still enslaves the American soul. In his book, *Democracy in Black,* Glaude's words are a gut check to those of us who broke out of the gates of marginalization, daring and hoping to grab a chunk of the American dream. In my defense, I didn't understand the psyche of colonizers and the chaos left in their wake. I chased my dream throughout the corridors of engineering and science institutions with only self-determination and very little desire nor encouragement to discover who I was, and where I was, or even "why" I was in the places I had been given access to. I had been dealt a disservice, I thought, but it still would not have changed the outcome of my tragedy.

One fallacy I once embraced, that I now know is a shallow promise, is the notion that education and hard work create limitless opportunities. My perspectives on the huge campaign to encourage young girls to pursue careers in Science, Technology, Engineering, and Math (STEM) has taken on a different and controversial twist. After 30 years in some of the most excluded places within the defense and homeland security domains, traveling to worldwide areas of civil unrest, designing and integrating military platforms, managing research on missile defense and directed energy systems, owning an engineering consulting firm, supporting the Chief of Naval Operations in the Pentagon, and integrating surveillance systems on aircraft and surface ships, I concluded that systemic inequities and racism cannot be resolved by simply equipping young people with science and math skills. It should not be the responsibility of aspiring engineers or scientists to eradicate disparities in the workforce which are not a result of diminishing talent, but are directly attributed to institutional racism. Despite terabytes of data generated to assess workplace diversity in the federal government, elaborate conferences to heighten the sensitivities of their federal managers, and baseless inclusion campaigns, the reality is that disparity and discrimination still exists within the corridors, cubicles, and conference rooms.

In a 2015 survey conducted by FedSmith.com, federal workers were asked if they witnessed racial tension in the workplace. Fifty-five percent of the respondents reported instances of racial tension directly, while 60% knew personally of someone who had been discriminated against. Eighty percent of the survey respondents felt the federal government's Equal Employment Offices were incapable of eliminating racism in the federal government.[25] My experience as a civil service employee, contractor, and consultant would substantiate that there is a pervasive culture of systemic racism, particularly in senior level positions, that extends beyond the civil service institution and erodes the fabric of the entire federal government network. This includes the workspaces of federal government contractors who are shoveled billions of taxpayer dollars in defense of the nation.

Riding the wave of maritime strategies has been both a blessing and a curse to my professional career. The blessing was that I was able to reinvent myself as the Navy adapted to a changing threat environment. I rode the technological waves of military modernization to transition my skill sets from a civil service scientist to a senior analyst in the Pentagon, but I also had more wipeouts than I anticipated. I landed in the Pentagon just as China's economic growth and regional relationships and its surging military spending and modernization were indicative of a security dilemma in the Asia-Pacific region. It was a post-9/11 era where the U.S. was in a conundrum with how to posture for nonstate actors, both homegrown and foreign. Iran had adopted a tactic of deploying hundreds of speed boats, a technique known as "swarming," in an effort to overwhelm U.S. platforms in the Persian Gulf.

Like most people, apart from Barbara Starr reporting from the Pentagon on CNN, I really didn't know how Washington worked and what took place in the Pentagon. I had encountered reductions in funding on contracts and I knew that in some way, there were people and institutions in the five-sided building that controlled the budget for the lower echelons of the Department of Defense. But I didn't have the big picture. I leveraged my past experiences to snag a job as an analyst with Cydecor, a small disabled veteran-owned business. I was assigned to work in the Navy Irregular Warfare Office (NIWO) directly supporting the Chief of Naval Operations (CNO). To synthesize the good, the bad, and the ugly of my experience in the Pentagon could take an anthology of writings. The maritime domain was posturing to address global threats such as violent extremism like those carried out by Osama Bin Laden's Taliban and Al Qaeda affiliates. I did not know when I applied for the job that the Navy's vision for confronting irregular challenges evolved from the *Cooperative Strategy*, but it did influence the jobs I applied to.

The U.S. Navy will meet irregular challenges through a flexible, agile, and broad array of multimission capabilities. We will emphasize Cooperative Security as part of a comprehensive government approach to mitigate the causes of insecurity and instability. We will operate in and from the maritime domain with joint and international partners to enhance regional security and stability, and to dissuade, deter, and when necessary, defeat irregular threats.[26]

The Navy Irregular Warfare Office (NIWO) was engaged globally to confront terrorists, assist in humanitarian relief, and build partnership capacity. This included efforts to bring stability in the Horn of Africa, where piracy, Al Qaeda affiliates, Al Shabaab, and the lack of a central government had increased the probability that the region would be a formidable security threat. I was assigned to assist my uniformed counterparts in identifying and fielding innovative technologies to support the warfighter in these efforts. Specifically, I was tasked with collaborating throughout the echelons of the Navy organization to identify research efforts and technology demonstrations at varying levels of maturity that could be rapidly deployed in maritime areas of operation. The Navy's vision for confronting irregular challenges was defined. Several efforts ensued to operationalize that vision and define measures of success to organize, train, and equip the troops to carry out the vision. It is impossible to really describe the complex relationship Cydecor had with the government without laying out the CNO's Operations, Plans, and Strategy organization as it was at that time.

The Navy Irregular Warfare Office (NIWO) was stood up by the Chief of Naval Operations (CNO) within the Operations, Plans, and Strategy department. The CNO designated Rear Admiral Mark Kenny, a Naval Academy graduate and submarine commander, as the first department director for the Navy Irregular Warfare Office. Admiral Kenny was responsible for the company I worked for securing a federal government contract to support NIWO functions. Contractors supporting Admiral Kenny's staff in the Pentagon included Cydecor's CEO, Nader Elguindi, who had been a young ensign on the USS *Birmingham* commanded by Admiral Kenny when Nader tragically lost his legs in an unfortunate motorcycle accident. Retired Navy Commander Dan Shinego, another Naval Academy graduate, was the Program Manager. However, Nader ran the show and asserted his authority for the eight to ten Cydecor employees on NIWO staff.

This ménage à trois between Admiral Kenny, Nader, and Dan Shinego would seem quite innocent except for the fact that Nader was the President and Chief Executive Officer of the company and was also responsible for forecasting, planning, and obtaining funding for all NIWO activities including those funds that were being disbursed to his own company! Well-compensated employees were willing to tolerate Nader's impulsivity, manipulation, and narcissism, but employee turnover in Cydecor's corporate office was high where more administrative employees worked.

Nader dragged his prosthetic legs throughout the Pentagon with a water bottle in one hand and flimsy backpack on his shoulders. Much of his manipulation came in the form of lavish employee retreats, invitations to his private boat, and lavish holiday parties. It was not unusual to have a young woman with little to no experience show up to work at the Pentagon as an analyst, totally oblivious to anything remotely related to DoD, the Navy, or defense acquisition.

My initial assignment was to develop a process for NIWO to transition capabilities and technologies to the warfighter in response to urgent needs. The Defense Department process for designing and fielding capabilities has always been criticized for the bureaucratic stovepipes and burdensome processes that create unwieldy and excessively long time frames for getting capabilities to the warfighter. NIWO, by direction of the CNO, had the authority to provide small amounts of resourcing to develop mature technologies. U.S. Navy Action Officers (AO) assigned to NIWO acted as advocates to match urgent needs and mature capabilities under consideration and to seek additional funding from other resource sponsors internal to the Navy as well as with other joint services to rapidly deploy capabilities for use in the maritime domain. Action Officers were naval officers who were temporarily assigned to support NIWO.

Prior to my assignment in the branch, the lack of a repeatable process resulted in the transition of technologies being highly correlated to the personalities and experience levels of the Action Officers as well as the length of the AO's tour of duty in the Pentagon. As AOs left the program, the requisite knowledge transfer went along with them, leaving their replacement to sift through stacks of briefings to figure out how to proceed with the transition. I developed the "Navy Irregular Warfare Innovations Methodology" for identifying, planning, resourcing, and rapidly demonstrating and fielding mature innovative capabilities. With the methodology, incoming AOs could track the capability from the initial need identification through resourcing the solution, to testing and fielding. I prepared the written plan and all the graphics necessary to determine technology transition readiness.

After several months of interviewing officers and drafting the plan, I received an email from Nader requesting that I write a justification for him to allocate NIWO funds to a large business defense contractor within Nader's illicit defense contractor clan, to develop the innovations methodology. Nader and I exchanged emails as I was trying to understand exactly what he envisioned the requirements were for the additional contractor support so that I could draft the justification and a statement of work. Nader explained to me that the additional support was needed to develop a process for NIWO to transition capabilities and technologies to the warfighter in response to urgent needs. This included establishing a working group to define the existing process for transitioning capabilities, conduct weekly briefs with the department head, Admiral Philip Green (Admiral Kenny had since retired from military service and reappeared on Cydecor's list of "advisors"), and then write the Innovations Methodology.

I asked Nader to expound on the requirement by distinguishing how the additional contractor support was different from the Innovations Methodology that I had previously submitted as an 80 percent-complete solution. Nader sensed that I was not going to blindly oblige his request, so he drafted a justification himself to authorize the additional contractor support, along with two statements of work, one to develop the methodology with an estimated contract value of $300,000 and a second statement of work for a consultant for an estimated $80,000 level of effort.

Nader sent the draft contract documents to me and again

requested that I provide a justification and draft statement of work. One of the things Nader underestimated about me was that I was fully prepared to forfeit my salary of more than $160,000 before I would do anything remotely considered unethical, illegal, or in violation of Federal Acquisition Regulations (FAR). This prompted the first of many "Come to Jesus" meetings that Nader and I would have.

We met at the 5th Floor, Apex 5 Subway restaurant in the Pentagon. I explained to Nader that I was refusing to follow his direction. My words were, "What you are asking me to do smells fishy and I will not be the one holding the fish when the auditors show up."

One thing that was obvious to anyone who had to deal with Nader was that he was terrible at containing his temper. He became visibly uncomfortable and was annoyed by my insubordination. His first approach was to insult me with patronizing diplomacy. He complimented me on the outstanding work I had done on the Innovations Methodology. He stated that the additional contractor would take the work I had completed and "put some skin on it." To which I asked, "What kind of skin do you want? I can put skin on it."

He stuttered, "Wh-Wh-What are you trying to say? I'm having a hard time reading you, Joy." To which I responded, "It's not that complicated, Nader. I've been VERY clear. READ MY LIPS. I'M NOT DOING THIS."

Nader grabbed his bag and water bottle and limped away briskly.

Nader had already scheduled working group meetings for the new contractors to present their progress to Admiral Greene. The only thing the new contractor lacked was the document that I continued to develop, which was ready for the Admiral to review, complete with "skin."

For several weeks, the contractors showed up at the Pentagon and laid out an approach for developing the methodology, which was merely a regurgitation of what I had told them I was doing. I cooperated by providing them updates on the actions I had taken each week. The contractor's representatives would come in and brief the Admiral as if they were doing the work, still without a signed contract. My guess is that Nader made two assumptions. One, Nader assumed that I would continue cooperating quietly and eventually surrender the methodology. Perhaps Nader was working on another willing victim to carry out his scheme. Nevertheless, when I was confident that the methodology was ready for the Admiral's final review, I presented him with the completed draft of the Navy Irregular Warfare Methodology, skin and all, fully aware that he would question the circus Nader paraded before him each week.

The Admiral did not waste time addressing the dog and pony show Nader promised him was forthcoming, where he would get a review on the methodology. I'm not sure how Nader explained it to the Admiral, but he allowed the contractors to come into the Pentagon, as they had for the previous three or four weeks, prepared to give the Admiral a status of their progress on the methodology. Nader invited the gentlemen and me into a small office on the fifth floor of the Pentagon for the purpose of having a private brief of their progress. As the men discussed the work they had completed (my work actually), Nader turned his chair and turned his back to them. I was totally not prepared for Nader's infantile behavior and neither were they. The men were forced to address only me as they completed their status, all the while we looked at each other in total confusion. Eventually, Nader interrupted the brief and proceeded to berate the men for their poor performance executing the project. Total schizophrenic behavior! Nader's behavior was so embarrassing, I somehow felt responsible. I called my counterparts the following day and said, "If Nader were my child, I would apologize for his behavior, but since he is my boss and the CEO of the company, I have no explanation for his behavior."

The Navy Irregular Warfare Innovations Methodology was soon adopted as one of NIWO's foundational documents and became a valuable process for Action Officers. Needless to say, my relationship with Nader became strained. I knew it was a matter of time before I would experience the wrath of Nader's revenge. I didn't know how and I didn't know when, but I was resolved to ride the bandwagon until the wheels fell off, giving 100% and exceeding the goals and objectives set for me.

A young lady appeared in the office one day. She was smart and attractive but she had no prior experience with the federal government, defense contractors, or any of the other military services. She was a recent college graduate and had been a nanny for someone within Nader's clan. For whatever reason, her previous employer terminated her employment as a nanny. Initially, Nader transitioned his resource responsibilities to her, which was a good place to start since she had a degree in business. To the young woman's credit, she really had no idea what she had gotten herself into, especially when Nader announced that she would be replacing me as the new Innovations Lead and I would be reporting to her. Poor baby. Her success depended on me telling her what to do and that was not going to happen. I knew, with certainty, my days were numbered after that.

Soon after, I was included on an email from Nader directing employees to charge our time to a charge code different from the one we had been using previously. The method contractors use to bill the government for their services is to create a charge code for each customer and each contract. Mischarging time is a felony offense and many contractors have suffered dire consequences because of such malfeasance. Nader had created such a bullying work environment, that few people would ever challenge him. But having been a small business owner and having endured several audits by the Defense Contract Audit Agency (DCAA), I immediately researched the existing charge code and the code Nader was directing us to charge to and discovered that they were for two separate contracts, two separate customers. One charge code was rightly allocated to the NIWO contract, the other charge code was for a contract that another Navy contractor was awarded. I was not doing work associated with that contractor or that Navy customer. I recognized that Nader signed my check, but again I was not willing to do anything unethical, especially something that could implicate me in government contract fraud. I simply responded to Nader's email by replying to everyone on the distribution and asked if the previous charge code had changed. Nader responded that the new code was to dissolve the previous years' funding for the same contract. Once again, Nader's fish was resurfacing. I accessed the databases I still had privileges to and found out what customer was really being charged for the work we were doing. Nader may have suspected I was smarter than he gave me credit for, so he followed up with a request for every employee to provide two weekly reports, one for the real customer, NIWO, and one for another customer whom I had no knowledge of. I continued to provide a weekly activity report for NIWO, but again, I refused to get tangled in Nader's web of fraud and deception.

Federal government contractors are no strangers to fraud

and deception. The very slick ones have a handful of military and government employees in their clan who are willing to generate bogus contracting vehicles in exchange for second careers after their first retirements. Veteran military officers and former Senior Executive Service (SES) federal workers retire on Friday and come back to their same jobs on Monday, employed by the same contractors that they were responsible for managing in their former positions. This all in spite of the statutory and regulatory restrictions concerning standards of conduct between contractors and government employees, which govern these unethical practices, conflicts of interests, unfair advantages in post-government employment, and misuse of government positions. For those who are squeamish about breaking the law or even the appearance of impropriety, contractors will pass them on to another company within the clan. Then there is the nebulous role of advisor that contractors use to increase their chances of gaining contracts. Contractors pad their profiles with the idea that they can "reach back" to former admirals and SESers for expertise in solving the government's problems.

The Office of Management and Budget (OMB) assists the President in managing the federal budget and overseeing the regulatory and statutory activities of the military and federal agencies. OMB will always be overwhelmed in the seldom-penetrated clan structures as found in the federal government contracting community. The Senate Armed Services Committee (SASC), which has legislative oversight of the activities of the Department of Defense, is constantly enacting policies to manage the share of nepotism and cronyism in Pentagon shenanigans. Prior to Senator John McCain uncovering the Boeing tanker scandal, few cases of officials performing favors in exchange for post-retirement jobs have been successfully prosecuted.[27][28] In the Boeing tanker scandal, a senior civil service employee, Darleen Druyun, used her influence in a contracting deal with Boeing in exchange for a high-paying job with Boeing after her retirement.[29] It is highly unlikely that the American taxpayer will ever know the extent to which their tax dollars are ensnared in the illicit transference of not just money, but wealth that circulates in the federal government contracting community.

If there was a saving grace that permitted me to remain in the Pentagon for as long as I did, it was the fact that after Admiral Greene retired, totally exhausted from trying to make sure Nader didn't involve him in his schemes, Admiral Sinclair Harris, an African American, assumed command of the Navy Irregular Warfare Office as its director. By the time Admiral Harris came aboard, NIWO had been forced to relieve Nader of his position as Resource Manager. The Defense Contract Management Agency (DCMA) had conducted a stand down to ensure NIWO was clear on the ethics of contracting and DCAA had done a surprise audit after suspicious charge accounting had been alleged. Nader's removal from NIWO had not averted the company's invisible hand into NIWO's budget, he just had to color inside the lines and stay in his lane.

For a brief while, I was able to make significant contributions without having to constantly look over my shoulder. I was responsible for the Assessment Line of Operations to validate the Air-Sea Battle (ASB) concept. In response to actors like China and Iran, who were gaining an edge on the U.S. in the development of weapons and capabilities that had the potential to deny U.S. military forces access to strategic geopolitical locations, the Navy and Air Force conceived a joint military strategy that integrated operations across all five domains (air, land, sea, space, and cyberspace) to create advantage.[30] I developed and implemented the Air-Sea Battle Integrated Assessment Plan, a process to facilitate the development of performance goals and measures, a Multi-Attribute Decision Analysis Process to compare measured performance with performance objectives, and the Air-Sea Battle Office Implementation Planning Guidance, which included the operational design for achieving the ASB end state and objectives.

On a day-to-day basis, I caught hell from the small remnant of officers who had not been transferred out of the Pentagon and were hanging around NIWO and were sucking up to Nader behind the scenes in hopes of securing post-retirement positions. Commander Bruce "Crash" Defibaugh thought I was ignorant to his role, that he had been appointed by Nader to turn up the heat on me in proxy since Nader had been removed from his position as Resource Manager. It was naïve of Crash to assume that I had made it to the Pentagon without having experienced most all of the tactics common in this environment. My presence and willingness to ask questions upsets the apple cart, interferes with people's bread and butter, and disrupts the status quo. Crash removed me from the Air Sea Battle team. I was reassigned to access open data sources to identify, track, and report on activity in the South China Sea and advances in Anti-Access/Area Denial (A2/AD) capabilities. I was informed that I would be reporting to the ex-nanny turned analyst in the Pentagon on matters of national defense.

The intensity involved in daily tracking the arms race between China and the U.S. and Iran's military modernization essentially made me unavailable to contribute anything else to the team. I had seen this before when the leadership team at NAVOCEANO quarantined me in the computer room to configure HP servers without clear requirements. I provided the same level of excellence toiling over open media sources, academic writings, and the spurts of information on social media just as diligently as I had when I set up HP servers to process hydrographic data. I followed Crash's orders and I was never delinquent on submitting the reports required and presenting them during the Admiral's weekly briefings. Fast forward several years, if you peruse the internet, you will find Crash, retired from the Navy and reporting to duty in the Pentagon as a civilian employee hired by Nader's Cydecor.

My experience in the Pentagon was mixed with joy and pain. I was fulfilled by the work I did, when I was allowed to do it. Nader, Crash, and others were successful at forcing me to control my urge to challenge unfairness and workplace hostility with all the vengeance I had inside of me. By then I understood that I was never going to win at the game. The system is not designed for those outside the clan. If I needed to cry, I went for a walk. And I prayed. I asked God every day, what was my purpose for being in the position I was in, in the Pentagon, just like all the other jobs I had been blessed to get, if I had to endure so much unfair and hostile treatment. My desire was to go to work, contribute to the team, and have meaningful workplace relationships, maybe even a close friend to hang out with after work. That has never been my experience. Going to work every morning required a lot of self-encouragement and focus on the work before me, while anticipating the next move of those hoping to catch me unaware and unprepared.

I prayed a lot during those four and one-half years in the Pentagon and I believe my prayers were answered. Due to the nature of the work in the Irregular Warfare Office, we often had to deal with asymmetric and novel tactics of war. Wargaming is a very often used strategy in the Pentagon, and I was engaged with a team of policy and strategy analysts responsible for mapping doctrine, organization, training, materiel, leadership and education, personnel, and facilities (DOTMLPF) to "kill chains," or the sequence of events for determining when to order an attack on a target. In one wargaming exercise the targets were nonstate actors, specifically Islamist extremists. John Sandoz, a former Navy Seal and a highly compensated consultant for the company, who was extremely cautious not to be jammed in Nader's web of deception and debauchery, seemed to have an understanding of Islamist ideologies that were critical in making decisions for identifying targets of interest and the complicated theological considerations for deciding when to strike. John was able to make the distinctions between Judeo-Christian and Islamist world views. I had a disturbing epiphany and a transformative revelation at the same time working alongside John. I knew absolutely nothing about Christian theology, despite claiming Christianity as my religion of choice, a practice passed down to me by my parents and distorted by all the different denominations I had subscribed to in my younger life. John stirred an insatiable desire in me that I didn't initially understand.

I became consumed with a desire to know more about all the Abrahamic faiths, Judaism, Christianity, and Islamism, but I especially wanted to know more about the holy book that I claimed to believe in, the Bible. I had an overwhelming, undeniable urge deep in my gut that would not let me sleep at night. I had visions of myself standing before audiences and I wrestled to wake from the vision of me exhorting. When I did wake up, I begged God to let me have a peaceful night's rest without the recurring dream of me speaking to attentive listeners. I am a hopeless introvert with extrovert skills. I had never dreamed of being a preacher. In fact, I struggled with some of the stories in the Bible. I was an engineer,

I told God. I wasn't sure how to make the leap from engineer to preacher. How would I pay my bills? I didn't even want to entertain the thought of preaching, but I found myself researching Christian theology. The first website that appeared on my computer was Virginia Union University, Samuel Dewitt Proctor School of Theology (STVU). Even when I tried to trick my fate by looking for doctoral studies in engineering, my computer seemed to have a mind of its own and would reload Virginia Union University.

Without exception, every single day that I walked on the campus of Virginia Union University matriculating toward yet another degree, my life was transformed over and over. This time I was fulfilling requirements for a Master of Theology degree. Dr. Yung-Suk Kim taught the first class I took and from the very first day, Dr. Kim assuaged my fears of questioning the Bible. With a better understanding of Christian theology, I must have become overzealous in my efforts to absorb all of John Sandoz's understanding of other faiths as well, or it could have been pure divine intervention that I received an email from John:

On behalf of the ICRD, Gabi and I would like to invite you to be our guests at the ICRD Faith-in-Action Dinner, Friday May 30th. This annual event is an opportunity to hear Congressman Michael Rogers' perspectives on U.S. national security challenges and learn firsthand about the ICRD's important faith-based diplomacy in addressing some of those challenges.

The International Center for Religion and Diplomacy (ICRD) Faith-in-Action dinner was held at the Mount Vernon mansion of H.E. Rafat "Ray" and Shaista Mahmood. Mr. Mahmood was the former Ambassador-at-Large for Pakistan. He and his wife came from Pakistan with almost nothing but a desire to help mend the ties between their native country, Pakistan, and the U.S. The Mahmoods own the largest house in the wealthy Mount Vernon neighborhood and they host the Faith-in-Action dinner annually. My daughter and I tried to contain our star-struck behavior when we left my black Suburban with the valet, ascended the majestic steps up to the main portion of the mansion and stepped outside to view the beauty of the Potomac while sipping wine, partaking in hors d'oeuvres, and meeting new and interesting people with interest or investment into the mission of the ICRD.

The ICRD has developed a model for transforming conflict in areas where there is intractable conflict by making faith part of the solution.[31] ICRD holds official consultative status with the United Nations Economic and Social Council (ECOSOC) and has used their approach to bridge religious considerations to promote world peace. ICRD has worked throughout the Middle East and Africa, Asia, Europe, and the Americas, bridging the gap between very hostile ideological and ethnic divides using methods they have matured that are not considered in the typical diplomatic toolkit. Yes, I was in awe at the level of wealth, knowledge, and experience in the room and the fact that I was there with my daughter who had just recently graduated from Georgetown's School of Foreign Service. And there was one speaker who captured my attention and rattled my inner stability, causing an insatiable unrest in a way that I would soon come to recognize as the language in which a higher power communicates with me. Rev. Canon Brian Cox, an ordained Episcopal Priest from Santa Barbara, California, and senior official at the ICRD, spoke at the dinner. It was as if my purpose for being in the Pentagon, my connection with John Sandoz, and the odd chance that I would be dining with the wealthy intelligentsia converged at that moment in time.

Rev. Cox is a pioneer and practitioner in integrating faith and politics in the international context. He has authored a book on faith-based reconciliation out of his experiences and methodologies he created working in Africa, Asia, Europe, Latin America, and the Middle East. In his book, *Faith-Based Reconciliation: A Moral Vision That Transforms People and Societies*, Rev. Cox emphasizes the significance that various moral visions have had in history and on the common good of humanity.[32] Even though I was in my second year of seminary, I was still struggling with my call. I was still not certain God wanted me to simply stand before an audience and tell the wonderful stories of my Christian tradition. I had already become a little jaded with the church as an institution for its apathy regarding social justice issues. Up until that point, seminary had been more of a self-discovery, a place where I was able to identify and articulate my own theology. I was beginning to heal from some inherently generational dysfunction as well as some induced dysfunction that I had normalized throughout my adult life. But there was still something unsettled. I had dragged my "otherness" for over 25 interesting years throughout the Department of Defense and Homeland Security and the results had always been the same, leaving me cast aside for not conforming to norms that wanted to find a "place" for me in order to satisfy a quota or to look compliant with diversity regulations. I was working in the Pentagon at the time, absorbing discussions centered on counterterrorism and building partnership capacity with weak, fragile, or failed states. I was perplexed that the very perpetrators who create the conditions of ethnic conflict and failed diplomacy were the same people who were focused on a resolution!

Rev. Cox described his "awakening" at realizing that somehow the faith-based community had to insert itself into informing, shaping, and influencing national security. President Obama's political theology challenged all faiths to "be willing to sublimate their ultimate theological and religious convictions for the common collective good" while urging the secular world to adopt a similar approach towards faith-based communities and their theologies.[33] I had my awakening that day. My life experiences had to somehow inform, shape, and influence national security. I have a perspective unlike most who have set the norms in the international community. My theological formation was just the beginning. I had to immerse myself into understanding the international community and the national security enterprise. I also still held a special place in my heart for the people of Somalia as that country had spiraled into anarchy.

My tour in the Pentagon was good while it lasted. One month after the ICRD Faith-in-Action dinner, I was the only person to receive a reduction-in-force letter terminating my employment. This was the true turning point in my life. My childhood dreams began to recur. In my dreams, I was flying. In life, I was just sailing through.

NO LONGER BOUND

I began my career as a civil servant on the hills of the notorious Watergate scandal. Republican President Richard Nixon's approach to civil service reform had created the model for future administrations. President Nixon was implicated in a cover-up of political operatives breaking into the offices of the Democratic National Committee, as well as other abuse of power charges. Although Nixon's eventual impeachment rippled throughout the federal government institutions, Nixon's strategy of structural change, reorganization, and increasing the number of political appointees prevailed in subsequent administrations.[34] The American public lost faith in the government. However, disregard for federal employees goes back as far as 1947 when President Truman issued an executive order to conduct "loyalty checks" of federal workers. Truman established "loyalty boards" within every department within the federal government. The Federal Bureau of Investigation (FBI) reviewed records and interviewed federal workers and categorized them as "totalitarian," "fascist," "communist," or "subversive." If a federal worker was found to be "disloyal," they were terminated. Although that system was eventually dismantled, the stigma federal workers carry has continued.[35]

Unfair competition within the Civil Service is especially prevalent in STEM professions. In an effort to attract the best and brightest talent, the federal government attempted to reform the civil service institution. The Civil Service Reform Act of 1978 (CSRA), which institutionalized the Civil Service System in response to a corrupt system of politics in the United States, was an attempt to dismantle the institutionalized disparity that associated federal employment with political affiliation, an effort that the Pendleton Civil Service Reform Act of 1833 had failed to achieve. Civil service workers have always been required to execute the political objectives of the current administration. CSRA required federal employment to be based on merit and by competitive examination.[36] CSRA also established a merit pay system for managers on the upper end of the General Schedule pay scale. Congress replaced CSRA in 1984, but the Office of Personnel Management (OPM) was still ineffective at enforcing the rules. The merit system is rife with ineffective management, political coercion, and disparity particularly for mid-to-senior-level federal workers, but especially for minority employees who are systematically devalued in the workforce.

President Jimmy Carter adopted Nixon's administrative presidency approach to civil service management in his State of the Union address in 1978. However, high on Carter's successor, President Ronald Reagan's policy agenda, was to cut the size and cost of government.[37] On the seventh day of President George H.W. Bush's presidency, he reminded federal government employees of their commitment to the political agenda of the office of President.[38] President Bill Clinton's response to civil service reform was to create labor-management "partnerships" which did not change disparities in the federal government. The tragedy of the September 11 attacks derailed radical attempts by President George W. Bush and were met with stiff resistance from federal employee unions and their supporters in Congress.[39]

During the administration of President Barack Obama, the federal government saw an increase in employee ranks; however, those numbers decreased as the 40th anniversary of CSRA was commemorated when President Donald Trump pushed to make good on a 2016 campaign promise to "drain the swamp." Trump, who repeatedly expressed outright distrust of federal government workers, took another stab at civil service reform by establishing a $1 billion interagency workforce fund to address things like hiring practices, disciplinary policies, job classification, advancement, and a lack of positive rewards for good performance.[40] [41] In another irrational move typical of Trump, he canceled pay raises for federal workers, claiming raises would strain the federal budget,[42] despite the fact that the cost to protect the billionaire Trump was unprecedented, exceeding the cost of protecting any other President.

Trump's swamp drain was indiscriminate against those he perceived to be disloyal. FBI Director James Comey was dismissed from his position when Trump accused Comey of being disloyal. Comey was responsible for an investigation into allegations that the Trump campaign colluded with Russian officials to try to sway the 2016 election, including utilizing stolen emails from the Democratic National Committee and Clinton campaign computer systems. Trump ran against Democratic nominee and former Secretary of State, Hillary Clinton, in that election. Trump revoked security clearances for previous high-ranking officials in his administration, indicative of his narcissistic demand for loyalty. Trump's administration perpetuated a climate of distrust and retribution that I also experienced more than 20 years earlier. The pattern of discrediting anyone who dares to speak out against irresponsible and unfair practices by an administration will always be the status quo.

The overwhelming need and bipartisan support for civil service reform contradicts the Federal Employee Viewpoint Survey (FEVS) published by OPM. FEVS measures employees' perceptions of their work experiences, their agency, and leadership.[43] It wasn't until 2002 that OPM administered the first survey of federal employees to get these first-hand, real people, real stories that give a snapshot of the federal workforce climate. It wasn't until 2016 that the U.S. Navy published a framework to strengthen its civilian workforce.[44] Federal government employees have never been fairly credited for the contributions they have made to the defense of this country. The civilian workforce is often overshadowed by dress uniforms and gold braids. The work of the civilian has uncharacteristically been seen as "easy," a "boondoggle," unattractive, just another "good government job." The civilian workforce is treated like stepchildren by the itinerant service members and political appointees, even though it is the civil servant who provides consistency and the transfer of knowledge. Federal employees have been unjustifiably mischaracterized as lazy and unambitious, content with the comfort of a secure "guvment" job, and for Black federal workers these stereotypes are magnified. According to a Federal News Radio Report in 2018, approximately 63% of federal employees were white and 80% of those federal workers lived and worked outside of the Washington, D.C., metropolitan area[45] in locales where attitudes and distorted ideologies are more likely to be displayed in the workforce, especially in the Bible Belt. Black federal employees have become accustomed to a lack of respect for fear of losing their jobs and the modest quality of living the jobs afford them. The prevailing trend is that the best and brightest workers leave the civil service workforce for higher salaries and more challenging and innovative responsibilities in search of reciprocal workplace respect. My experience has been these three ideals are not mutually inclusive. I have had good salaries and total disrespect. I have been in charge of innovative development of military capabilities, while enduring unbelievable and intentional disrespect as well as passive aggressive hostile work environments. It is not unusual for older workers who are former federal employees to return to federal service years later when they care more about working towards retirement than how fair the workplace is. Under the federal government's reinstatement eligibility process, previous federal employees may be rehired without competing with the general public.

Even though CSRA gave federal workers a voice and a vehicle to form a union and petition the government, potential and existing federal employees have few options for addressing retributive actions or unequal hiring and promotion practices, particularly the types of discrimination seen in professional job classifications. There are few federal government organizations that advocate for the federal worker, and none of them have the power to change the federal government culture. The American Federation of Government Employees (AFGE) is a labor union which grew out of the hardships federal workers saw during the Great Depression and has grown to a self-serving organization, the largest federal employee union representing 700,000 people that includes federal and D.C. government workers nationwide and overseas. Because of the geographic and institutional breadth AFGE has taken on, the impact of AFGE's efforts is hardly measurable on a department or agency level.

Blacks in Government (BIG), the initial vision of a small contingent of Black civil service employees at the Public Health Services (now known as the Department of Health, Education and Welfare), exists to eliminate practices of racism and racial discrimination against Blacks in government. The most effective goal BIG has accomplished is advocating for career training for employees, but the organization has not been a commanding force in integrating the decision-maker space of government.

The federal government and the contractors that support it will always be a squeaky machine. The institution works, but there are a lot of holes in it. There are rusty spots amongst the shiny places. There are just enough empty spaces to shovel in the obvious disparities. It would be delusional for the federal government and its contractors to believe that the ethnic, racial, cultural, and generational tensions that play out in the community, the media, and in social platforms have not spilled over into the workplace. There is just enough policy within this space to mediate conflict and prevent mayhem in the workspace. But there are those, like me, who harbor deep resentment for unmitigated injustices. I'm pretty sure there is an underground network where my name is coded to represent and forewarn employers of my inquisitiveness, determination, assertiveness, and unfaltering search for excellence and equality. I have grown and matured. I no longer have a desire to chase the illusive American dream.

My experience as a civilian serving this country had high moments where I had great anticipation of making a significant contribution to defense of the nation. For every high, there was a low shrouded by some person, some situation, some agency, some policy or status quo that did not value me or my contribution. Any flicker of hope that there was still a chance was doused by four simple words, "*Make America Great Again.*" Those four words independent of context are constructs that conform to the rules and conventions of language. But the political hostilities of a potential and the eventual inauguration of a Trump administration marked divisive sentiments around the country. The rise of distorted nationalists' ideologies in America was exaggerated by social, ethnic, and political unrest and the tensions trickled into the corridors of workspaces and schools, as well as within the public square. American nationalism contorted into something that was "dangerously irrational, surplus, and alien."[46] Black men, Black boys, and Black women were being harassed and killed by everyday citizens and rogue police alike, with no arrests or consequences. Restrictive nationalists made up a large portion of Americans and accounted for more than 30 percent of survey respondents where Bonikowski and DiMaggio identified four clusters of American ideologies in 2016, each characterized by a distinct nationalist disposition. The ideologies of restrictive nationalists believe that to be "truly American" one had to be Christian, speak English, and be born in the United States. I would add to that assessment, "and white."

In a country once revered for welcoming the poor and huddled masses onto its soil, a Trump administration seeks to fortify and seal off America's borders. At the top of Trump's campaign promises was building a wall across the country's southwest border to deter illegal immigration. Children were separated from their parents at the borders to discourage others from attempting to enter the country. Trump enacted two controversial Executive Orders (EO). EO 13767 enacted "all lawful means to secure the nation's southern border with Mexico, to prevent further illegal immigration into the United States, and to repatriate illegal aliens swiftly, consistently, and humanely."[47] Executive Order 13769 halted all refugee admissions and severely restricted travel to the U.S. by people from seven Muslim-majority countries—Iraq, Syria, Iran, Libya, Somalia, Sudan, and Yemen. The order, titled "Protecting the Nation from Foreign Terrorist Entry into the United States," introduced a cap of 50,000 refugees to be allowed into the United States in 2017, about half the number set by former President Barack Obama.[48]

By the time Trump began to "drain the swamp," my career with the U.S. Navy as a civil servant, contractor, and consultant had spanned nearly 30 years. U.S. maritime strategy played a significant role in my rise as well as in my fall. For every victory, there was an accompanying defeat. *From the Sea* had directed a path for me to navigate my career. I followed the currents forward from the sea and I endured the pitch and roll of some very turbulent waters in my career aspirations and I was well battered from the journey. I had been trained in both land and sea navigation. Dead reckoning (or DR) is the process of calculating one's current position by using a previously determined position and advancing that position based upon known or estimated speeds over elapsed time and course. The term "dead reckoning" is also used to refer to the process of estimating the value of something by using an earlier value and adding whatever changes have occurred in the meantime. Dead reckoning is subject to cumulative errors. By analogy, for nearly 30 years of my life I had been positioning myself using the Navy's maritime strategy and advancing my position based on my experiences and mistakes in previous positions. I had a strong desire to loose myself from the cycle that repeatedly found me lost and off course by dead reckoning. My ship had run aground and I was holding on to broken pieces trying to survive. I finally resolved once and for all that I "did not want to study war no mo," at least not in the same capacity that I had for the past 30 years.

I was in my final year of seminary when I was released from the Pentagon, so I focused on finishing my academic year strong, forging long-lasting relationships. Meanwhile, I picked up a couple of short-term contracts as a consultant. For several months I analyzed alternatives for various DHS, Transportation Security Administration (TSA), and commercial-as-a-service (cloud) solutions and made recommendations for TSA to align its mobility strategy and enterprise architecture with DHS's cloud-first initiatives. I was also able to support a contract with the Naval Air Systems Command on a part-time basis.

I spent a lot of time supporting my church during my unscheduled sabbatical. My pastor, who was also my homiletics professor, Dr. James Henry Harris, had become a tremendous help and mentor to me, even with his unconventional means. He had a reputation for being tough and inflexible, but his methods have transformed the vocation of preaching and preachers across the country. Dr. Harris helped me find my voice and how to use the written word to paint the thoughts that were in my head. Dr. Harris is more than a preacher, pastor, and teacher; he is an author, a scholar, a philosopher, an activist, and most of all, a friend. Dr. Harris, author of the book, *No Longer Bound*, often quotes the profundity of Dr. Martin Luther King, the morality of Immanuel Kant, the phenomenology of French philosopher Paul Ricoeur, and the existentialism of Jean Paul Sartre. So typical of Dr. Harris are signature quotes of his own, such as, "Reading is an act of holiness" and "All preachers lie." His sermons are part prophetic, part scholarly, part philosophical, and part comical. His teaching released me from my struggle with the call to "preach." He helped me realize I was not bound to the stereotypical preacher to perform as a cheerleader and abrogate the rights of people to do whatever it is they choose to do with their lives, but I had freedom and as Dr. Harris exhorted, "a responsibility" to interpret texts, many texts, and to interpret life as it is lived in the present. As Dr. Harris describes in *No Longer Bound*, I had a divine charge to make intertextuality and the surplus of meaning in life plain to all those who heard what I had to say, which contributes to the "eloquence" of the preacher.[49]

I never expected the federal government culture to spontaneously adapt to diversity in the workplace, but I'm not sure if I would have ever given myself permission to explore a different aspect of service to my country had it not been for my career experiences, which really culminated in the Pentagon. Admittedly, I already had my share of disappointments navigating the defense space, but I always held out hope that the next opportunity would somehow be different. During the time I was unemployed or underemployed, I was able to take a personal assessment of where I had been and how I managed to find myself in a place with so much education and experience and yet, I had no job to go to, let alone, excel in. I had no enthusiasm to offer an employer. I would not have hired myself during that long year of unemployment.

I had given all that I had to give, but I was in no position to give up. I still had to keep a roof over my head. I considered full-time ministry, but there was something about being limited and bound to the local church that did not motivate me. In fact, I am not really a supporter of the status quo culture of the local church. Dr. Harris and my church family at Second Baptist Church in

Richmond, Virginia, saw the church as a place for the community to gather, to feed the hungry, and converse about issues of social injustice, not only in the local community but throughout the nation. I didn't have a plan, and honestly, I didn't want one. For the first time in my adult life, I was no longer bound and I decided to go wherever the wind blew. I wanted nothing more than to distance myself from maritime strategy. I wanted to separate myself from the pseudo-identity I had created within the DoD and the Navy, realizing that I had never truly assimilated into the Navy culture and it was not for lack of trying. Throughout my career, I was a member of the National Defense Industrial Association (NDIA), the Association for Enterprise Integration (AFEI), the Defense Acquisition University Alumni Association, the Institute of Electrical and Electronics Engineers (IEEE), the Society of Naval Architects and Marine Engineers (SNAME), the International Council on Systems Engineering (INCOSE), the American Society of Naval Engineers (ASNE), and the Navy League, just to name a few organizations. I contributed more than I gained by associating with these organizations. I have traveled the country attending conferences, seminars, and symposiums, shelling out thousands of dollars in registration fees, travel, and incidentals. I did, however, gain knowledge, but the return on my investments has been negligible.

After 13 years on Navy ships and even longer designing naval platforms, I thought I was just as much a sailor as my uniformed counterparts. But time and time again, I was reminded that I was undervalued. I left the Pentagon wanting nothing to do with Navy culture. I cringed when I read my own résumé. I wanted to distance myself from who I had known for the past 30 years. I applied to entry level jobs, positions as an executive assistant, and even jobs in the hotel and restaurant business. I was willing to start over from the bottom.

After a year of unemployment, I found work supporting the Naval Air Systems Command H1 Helicopter program at Patuxent River (PAX River) Naval Air Station. I accepted an offer of a salary that was less than half of what I had made annually in the previous ten years. I lamented with the Black federal government employees who had the same complaints of unfair treatment and disparity as I experienced more than 20 years prior. They huddled outside in the parking lot to console and lick each other's wounds as they discovered what I already knew, and that is there is no recourse for complaining that ever works in the Black employee's favor. The culture was that federal managers emotionally whip Black employees into submission by hitting them with unsubstantiated performance deficiencies. You either shape up without complaining or ship out.

I wasn't interested in climbing any proverbial ladders of success. Hell, the truth was, the only reason I took the job was for the health benefits. The physical, spiritual, and emotional beating I took during my employment at PAX River caused me to question everything I thought I knew about myself, my strengths, and my purpose in life. The two and one-half hour commute each way to work and back quickly became unbearable, so I found myself shivering in the winter cold in the back of a coworker's farmhouse in the middle of Amish country. The generosity of a stranger was extended to me in the form of housing in a former railroad depot, converted into a one-room cabin. I was the only person of color for at least a 30-mile radius. The back of the cabin faced a wooded trail that could have easily facilitated my demise if anyone knew I was there and had intentions to do me harm. Not a night went by that my Ruger .357 magnum was more than an arm's length away, that my mind did not conjure up images of the 1969 incident when the Klan surrounded my family vehicle in Lake Charles, Louisiana. My experience camping in the African desert made sleeping in my coat, gloves, and hat a better alternative than driving two to three hours home and back each day. That was until the pipes in the cabin burst and I went without running water for a couple of weeks. Still, I was fortunate that due to the generosity of a coworker I found shelter in another rural, more "developed" living quarters. I am indebted to both of these individuals, because I was humbled that they allowed God to use them as instruments of God's grace toward me during a difficult transition in my life.

Left alone, confined to the sound of wind and the occasional barking and hissing of nearby animals, I escaped into a whirlwind of pity parties, wrestling matches with God, and eventual self-actualization. I survived by the grace of God. I was also taking doctoral classes towards my Ph.D. The world of academia helped me to breathe. I was interested in the intersections of faith and diplomacy, a desire that was divinely planted and watered by my experiences in the world of war and the weapons of warfare, both in defense of the country and my own survival in the culture of the federal space. My encounter with John Sandoz and the International Center for Faith and Diplomacy had pulled me out of a rut I was digging deeper and deeper into. I emerged from solitary confinement confident with my own strategy. The Master of Theology degree was just the first phase. I was on my way to the doctoral degree to legitimize my ability and qualifications to ask all the questions I wanted, and the only person I had to challenge was myself.

To add to all this turmoil, my Momma's health was failing. It had always been my prayer that I would be able to care for Momma as she began her journey back to our creator. Watching her condition deteriorate, I had a sudden inclination to search for a job. I didn't search for a particular location. I was not looking for a promotion or more money. I lost all aspirations for the coveted trophy, "Queen of the DoD." I had been defeated by that beast. I simply typed the words, "Ocean Engineer," into a search engine. With one click the first job that popped up was for a Chief Ocean Systems Engineer. The position could be based in either Arlington, Virginia, or Long Beach, Mississippi. I was interviewed for the Chief Ocean Systems Engineer position, but I was not deterred by the fact that I was hired as a Senior Systems Engineer. I didn't care about the title or the lower level of responsibility. Either way, the job was a win-win for me. My salary was restored to a level indicative of my years of experience, education, and security clearance. I was no longer sleeping in a stranger's cabin. And more importantly, I was able to see my Momma at least once a month.

The company that hired me was the crème de la crème among federal government contractors. What started as a small employee-owned company in 1969 grew in 50 years to a cyclopean family of divisional competencies, with over 30,000 geographically dispersed employees and ten other subsidiary companies. Through a series of strategic mergers, the government contractor secured an edge on the competition within several federal government domains, but particularly the DoD. The company's employees specialize in areas of cyber, IT, and systems engineering and integration of technologies from research and development projects to a number of land, sea, and air platforms and are considered the best and brightest. Employees work alongside the government and military in sensor and software development for some of DoD's most complex weapons systems. Everything about the company impressed me.

On the surface, the company had gotten everything right. Onboarding was efficient and thorough. I had a corner office and a working computer on the first day of work. Everything from time charging to cybersecurity to signing up for benefits was integrated into the company's IT infrastructure with a single sign-on. Organization charts, employee profiles, and the wealth of corporate knowledge was made available to employees. The company promotes its Employee Resource Groups (ERGs) for women, minorities, millennials, and the LGBTQ community. All employees are encouraged to participate in ERGs and communities of practice to engage in virtual dialogue, and to contribute both personally and professionally to a collaborative media platform. Employees receive points based on online contributions. The President addresses new hires in person and is often seen throughout the company and in the community promoting awareness of health and other social concerns.

The company set the standard for its effort to be inclusive and transparent, but I had no illusions that familiar spirits would not hover over the work spaces. I was the Senior Systems Engineer for an underwater surveillance system that was being integrated in Long Beach, Mississippi. I reported to the Program Manger in Arlington, Virginia, on a day-to-day basis, but I had direct reports in the Long Beach office. Bill was the Principle Systems Engineer and Alice was the Configuration Manager. Although Alice and I got off to a good start the first day we met, Bill's reluctance to report to me was evident from the beginning. Neither of them really deserve an honorable mention in the memoires of my life, except for the fact that they provide perfect examples and epitomize the two extremes of society that have for decades past and for decades to come will thwart all the good intentions of the federal government and its contractors to eradicate conflict in the workplace. There is no way to legislate stained hearts and minds.

But for God having mercy on Bill and blessing him with enough intelligence to escape generational poverty, the silence of the slightly pudgy, red haired, homely introvert spoke loudly of his resentment of my authority. Bill rarely, if ever, looked me in the eye. His subtle and deliberate passive aggression poorly masked his discomfort with my presence. He could not respond to a request I made without consulting my superiors. He had an aversion to any requests that came directly from me. He directed Alice to confer with him on anything I asked of her directly. I was neither surprised by, nor put off by, his sullenness, a familiar spirit I often encountered in the workplace. It has always been interesting to observe behaviors of people who assume they are covertly maneuvering the workplace, unaware that their tactics are not novel and are amusingly predictable! I came to the company with over 30 years of experience working within the federal government space. There was almost nothing I had not encountered. I say "almost" with much emphasis.

The Program Manager and Chief Systems Engineer were seemingly unbothered by my presence, after all I had interviewed with them, as well as with my supervisor before I was hired. I was informed early on that my authority over the Long Beach office would be met with hostilities and that I would be "moving people's cheese." I was briefed on Alice's incompetence when I arrived and I was challenged to un-muddle Alice's responsibilities and her unacceptable performance. I sought to approach this sensitive matter just as I do all sensitive issues—observe, document, and make gradual improvements to inefficiencies. I spent a considerable amount of time with the Program Manager and Chief Systems Engineer when the program was in its infancy. The government funded the proof of concept for the technology and it was our responsibility to transition the technology and deploy it to strategic areas of operation for the purpose of eventually becoming a program of record.

The Program Manager was undoubtedly one of the smartest women I have ever worked with. A hopeless workaholic, she was a former federal government manager with an impressive background leading several high-profile defense programs. I often worried if my life-work balance was as askew as hers. Her incomparable work ethic seemed to have cost her the joys of life. When we traveled, we talked about work on the way to the airport, in the airport terminal, on the plane, at Panera Bread between the airport and work, and then we worked all day. After work, we sampled great restaurants while discussing the day's work. We started each day in the hotel lobby strategizing for the upcoming work day. I believe she was aware that her life was imbalanced, but for whatever reason, work defined her.

The Chief Systems Engineer, on the other hand was a great resource for things concerning life and work. He was just as much academic as he was personal and professional. It was not unusual for us to grab a sandwich between meetings while chatting about attention deficit children or moves from the South to the mid-Atlantic. He was also a very good resource to ask questions. He had a Ph.D. in computer science and was a Technical Fellow, a status bestowed on a very small percent of technical staff. Technical Fellows consistently demonstrate expertise, leadership, and application in technical innovation. In 2017, the 42 Technical Fellows included five women, and if you look only at surnames, there is a possibility that two Asians and one Hispanic employee met the rigorous requirements. Technical Fellows are an elite lot.

For several months, I rode the wave of tight deadlines and intense requirements, standing up new processes, improving on others, and meeting the government's fluid demands. The Program Manager kept a tight rein on programmatic issues and I collaborated with the team on my progress managing the technical nuances of the program. My supervisor checked in periodically and gave very positive feedback on my performance and integration on the team. Bill continued to evade my authority and Alice reneged on all of her program responsibilities. Everybody avoided confrontation with Alice, who had a penchant for spewing venom. It would be easy to picture Alice as the other side of the "what's wrong with America" coin. Alice would be the one standing outside a dilapidated house, disheveled and disoriented, complaining about minorities ruining her life, constantly finding a scapegoat for her pitiable existence. The circumstances under which Alice was hired, along with a few other friends, brothers, and in-laws (the FBI) in the Mississippi office surfaced constantly in private conversations.

On one of my frequent visits to Long Beach, I walked into Alice's office to speak with her directly about missed deadlines. I had attempted to have the discussion by phone from Virginia, but Alice had abruptly hung up on me and sent me an email to let me know that she would not be talking to me. The Program Manager suggested that I go to Mississippi to "bond" with her. That was a new one, but hey, I would get to see my Momma. Alice chose Bill by proxy to justify why she could not submit documents that we were contractually required to provide the government. Because I was ultimately responsible for these submissions, I ended up doing Alice's work month after month. I kept a very detailed written account of Alice's shortcomings. I had been given latitude to right the wrongs of Alice's insubordination, or so I thought.

It was a typical nonstop flight to New Orleans, pit-stop at Panera Bread in Slidell, commute to Mississippi kind of day. I set up in the "hotel suite," a temporary working spot at the Long Beach facility and immediately commenced working as I had done on all other occasions in the Long Beach office. Bill expectedly closed his door to prevent interaction with me. Despite the fact that I was just around the corridor, he chose to cower in his office and propel emails to me through the atmosphere. I didn't care as long as the work was done. Bill may have resented my authority and avoided engaging me in conversation, but he was a conscientious worker. On this particular day, I walked into Alice's office. I was not in her office long enough to observe what she was doing because as soon as she looked up and realized I was in her office, she banged her fist on the desk and yelled, "FUCK!" Obviously, Bill had failed to warn her I was in town. Lucky for Alice, I valued my paycheck more than I did her broke-down, stressed-out trailer park mouth. I was kind enough to offer her a second chance at a good morning. I excused myself and told her I would step outside her door and re-enter and give her a chance to rethink what she had just done. Without listening to her rebuttal, I kindly walked out, stepped back three paces, and re-appeared alongside Alice's desk. Not much had changed in her demeanor. She informed me that she was trying to get her work done and that I was "bothering her."

When I realized that Alice's behavior would not be dealt with seriously and intentionally to subvert her unacceptable breakdowns, I knew this would be par for the course. Emotional outbursts by white women are typically ignored in the work place. I had to adapt once again to the status quo and I would be the one accused of inappropriate behavior, for having dared demand that Alice do her job and be reprimanded for insubordination and disrespect of my authority. Although I submitted a very detailed report of Alice's behavior to Human Resources, I soon began to feel the retribution of having dared expect Alice to do her job and respect me as a person.

Months went by and Alice did what Alice wanted to do when Alice wanted to do it. Bill continued his passive aggression. The Chief Systems Engineer transitioned off the program and for a couple of months, I filled in for those responsibilities. Even though I was in work mode more than 12 hours a day, I was rewarded in seeing my Momma's smiling face as she greeted me from her leather recliner every time I walked through the door. I knew my days with my Momma were numbered. The demons of systemic disparities, pathological ignorance, and racial intolerance were not strong enough to knock me off my square. My priority was squeezing every ounce of life out of what I had left with my Momma.

The new Chief Systems Engineer eventually showed up in the person of a Hispanic gentleman named Juan. Lord, I did not see that one coming! I thought I had seen it all in my 30 years interacting with the federal government and government contractors. But there is always a first time for everything. Juan met with the engineering team and allowed us to share the successes of the program as well as its hindrances. The government customer for the program had "strongly" recommended that the company hire a white lady from California, so that she would not spend months on the street looking for employment. The new hire could remain at her home in California while working remotely. She and I shared with Juan the complications we encountered working with Alice.

Juan interrupted our meeting only to return shortly. There was an obvious smothering shift in the atmosphere after Juan's break. When he returned to the conference room, Alice was with him.

A phenomenon I was all too familiar with began to take shape. I observed as Juan supported my theory that the biggest distinction between a poor manager and a good manager is that the poor manager seeks first to weaken the influence of the strongest personality on the team. This behavior is usually grounded in deep seated insecurities and strong desires to be accepted and recognized, usually after many failed attempts. The weak manager has to find a more submissive type, a good follower who would be all too willing to execute tasks the weak manager is either incapable of doing themselves or who knows just enough about ethics not to have their behavior impugned. I would wait to see who Juan chose, Bill or Alice. But I had my suspicions it would be Bill. He was far smarter than Juan. Alice was too inconsistent to be anybody's scapegoat.

I would not have spoken about the obvious shift in Juan's attitude toward me, had the coworker from California not mentioned it on our ride to Darwell's Café, a much-needed break from the tension that engulfed the conference room when Juan returned with Alice. My coworker warned me that the tension seemed directed at me and that I should be very careful around Juan in the coming weeks. This confirmed my suspicions.

I don't think the gravity of the insensitivities really hit until the next morning as the team sat around the conference table waiting for the start of yet another meeting. I sat quietly as my coworkers engaged in an interesting conversation about how much they missed Little Black Sambo. This was in the year 2016—in the United States of America! It took everything in me right down to the cartilage in my pinky toe not to erupt into the angry Black woman. I remember telling myself I needed my job. I needed to see my Momma. Even when someone remarked that it was a shame that people were so sensitive about "stuff like that," I held my peace. I wanted to cry, but I had long learned to recite the words written by Colonel Jo B. Rusin in her book, *Move to the Front: A Guide to Success for the Working Woman.* In Rusin's leadership secrets from a woman soldier, she advises, "Never let them see you cry," when dealing with emotions in the workplace, Rusin says, is a cornerstone for respect.[50] I don't know if I ever earned respect from my coworkers by not crying, but managing my emotions went a long way toward maintaining self-respect and that's all that matters to me at this stage in my life. If I had to point to a moment in time where I began to lose hope that things would be different, the Little Black Sambo conversation would be the pivotal moment.

A couple of days later, I arrived at the airport in New Orleans, eager to return home and regroup. I was surprised to see Juan in the terminal. I had not realized we would be on the same flight back to Washington, D.C. I sat up when I saw him entering the seating area near our gate because I assumed he would join me. Juan looked at me in disgust and walked past without acknowledging my presence. I was not broken. I was actually empowered. For someone who had just met me a couple of days earlier, Juan had given me more power than I really deserved in that moment.

In the weeks and months that followed the Little Black Sambo conversation, Juan shifted every project I was previously working on to other people. I'm sure I didn't help my cause when Juan called one day and said he wanted me to start a telephone log of people he contacted regularly. I suggested he choose Alice for that. Juan also blurted out in a weekly staff meeting that he would be getting me to do some of the administrative tasks he had been too swamped to get to. A simple, "Excuse me?" and a look of defiance was all that was necessary to squash that idea. Thirty years of experience and Rusin's recommendation to know office tactics and how to deal with bullies was essential to my survival under Juan's regime.

I accepted the reality that I was in a losing battle with the entire defense domain of politics and culture clashes. Juan's bullying changed the entire dynamics of the office. He assigned me to rearrange thousands of electronic files on the program's SharePoint site, something that took a couple of months to complete because of the many technical incompatibilities inherent in the infrastructure and the fact that Juan would repeatedly change how he wanted the files organized. We were no longer having weekly staff meetings. Juan held meetings in private conversations. I discovered in a briefing to the government that Juan reorganized the Systems Engineering team, removed Bill and Alice from working directly for me, and directed them to report to him. It became apparent then who Juan had chosen to be his scapegoat. Bill began to mirror Juan's disrespect and avoidance towards me. Juan gave Bill authority to make changes to the government's baseline design without going through the Configuration Review Board (CRB), which was under my purview. I made numerous requests for an explanation for unauthorized changes. After conducting a configuration audit, I finally refused to cover up any more configuration changes.

No one questioned why I went from being a highly productive member of the team to shuffling files around in cyberspace. I'm sure they spoke about my silence among themselves. No one questioned how I went from "doing it all" to doing "little of nothing" until... there's always that moment when we all have to deal with adversity head on whether we do it willingly or it is forced upon us.

That pivotal moment to deal with the situation that Juan was instigating was forced when the Program Manager included me on the short list of engineers to attend a supplier symposium. It was the first of its kind that allowed current and prospective companies to showcase innovations and collaborate with the company's "diverse set of leaders." By this time, communication with Juan was in total breakdown. I felt hopeless. Juan managed the team as if I didn't exist. I volunteered to assist with fault identification and failure analysis. The deployed systems were experiencing deteriorating operational availability. The government's concerns were escalating and we needed immediate attention to identifying root causes of the failures.

I arrived early at the Hyatt Regency where the symposium was being held on the evening of the event. The next morning I found a nice comfortable chair in the lobby of the well-appointed hotel. I had a paper to complete for a class I was taking. I enjoyed a cup of coffee and got an impressive amount of writing done for about 45 minutes. I moved into the ballroom about 30 minutes before the speakers were to start. I took a seat in the back of the ballroom. I was seated long before my coworkers arrived. As I sat there, I was struck by the obvious fact that of the nearly 500 people convening in the Hyatt to meet a "diverse set of leaders," the participants were nearly all white males. I had to remind myself that I no longer had the desire nor the bandwidth to disrupt the status quo. Neither was I in a position to do so. I had to dial down my aggravation at the fact that out of 30,000 employees, mostly engineers, scientists, mathematicians, and IT professionals, of all the diversity and inclusion meetings, the Employee Resource Groups, out of all of that, the company couldn't find three Black employees to invite to this symposium. Well, I think I did count three, but it wasn't clear if they were employees or suppliers.

As the room began to fill, the Program Manager spotted me in the back of the room reading through the paper I had just spent the last hour completing. She invited me up front to a table with the rest of the people from our team, including Juan. I would rather have stayed where I was and read my paper. I was not interested in socializing with people who thought it a shame that Little Black Sambo was banned from schools and roadsides across America. I had no desire to pretend we were a happy, cohesive team of professionals who treated one another with mutual respect. I was still trying to figure out where all the "diverse set of leaders" were that was touted for this symposium. The symposium had not yet begun and already I was in my feelings about how this day would progress. I was in a sea of predominantly white men, and Juan who didn't know the difference. Executive leaders and key speakers were taking their places on the stage, all white men. The few people I did know in the room had already demonstrated how little they valued my presence. I was still sulking over how Alice was allowed to be outright insubordinate and hurl profanities at me without reprimand, how I was excluded from meetings, and relegated to doing administrative work. Hell naw, I ain't trying to break bread with ya'll! But... I did. The Bible says that the Lord will prepare a table for me in the presence of my enemies. Maybe this was that table.

One thing that really boils my crawfish is when people take a snapshot in time, say for instance when a football player takes a knee instead of standing during the national anthem or when a 23-time Grand Slam tennis champion who just happens to be a Black woman passionately defends her integrity with a referee who accuses her of cheating in the U.S. Open. Snapshots are frames of a motion picture. Just like Alice's f-bomb was probably intrinsic to all the other chaos going on in her life, I gave her the benefit of the doubt and a chance to start her Monday all over again. Even with all this prior understanding of human proclivity to see everything out of context and to be more concerned with the "whys" rather than the "whats," I was not surprised when the Program Manager called a few days after the symposium with her concerns. Despite all the times I went to her for help, she quoted Juan. I was "not being a team player." Juan didn't feel my work was "reflective of a Senior Systems Engineer." And there it was. I expected it, which it why I kept a paper trail of all my conversations with Juan and the history of emails I sent asking him what his expectations were of me. I kept a journal of all of the things I accomplished before and after Juan arrived. I resurrected the email I sent Juan with my résumé requesting that he reconsider some of the administrative tasks he was assigning to me. I had also documented my encounters with Alice and sent them to Human Resources several months prior to this predictable assault on my performance.

Juan had nothing specifically performance-related to hang his accusations on. The Program Manager was more concerned that I didn't want to sit with them at the symposium. I went to speak with Juan to find out what I could do to turn the capsized ship upright again. This was nearly six months after Juan assumed the Chief Systems Engineer position and seven or eight months since Alice dropped the f-bomb. Sitting across from Juan in his office he asked me, "Did you ever apologize to Alice?" There's no place on this paper to insert googly eyes. I was done. For real. For real.

A couple of weeks later, I logged into my computer to start the day's work. There was an email from the Program Manager.

I was being removed from the program's funding. A week later, my Mom died.

I have no illusions that the discriminatory culture of the federal government and its tentacles in the private sector will one day evolve into utopian existence. While this discourse on race and the government is not a novel subject, this account of my experiences may somehow reverse the mirror and magnify beyond what can be seen in the data and statistics about workplace diversity the federal government and government contractors reveal. How the perpetrators of disparate, unfair treatment within the federal government space feel and how they react to these accounts will assuredly challenge their truth, which is grounded in their ideologies that are in opposition to my own.

When I walk into a room, I immediately see the one thing that is not like the other. Homogeneity screams at me and reveals a greater, deeper story than the possibility that happenstance created a same race, same gender, or same ideological view in the workplace. I have been incalcitrant to accept distorted interpretations of ancient texts, supreme deities, outdated rules and norms, and sanitized statistical data to rationalize acts that marginalize and disparage groups of people. I've been asked out (or kicked out) of places, including the church, because I dared to disrupt the status quo.

I grew up in the Navy culture. I learned when to challenge and when not to challenge as well as how to challenge. Verbal confrontations and emotional outbursts would never work in my favor, not that I considered having an emotional breakdown, even when I had every right to defend myself in the face of danger or disrespect in the workplace. Daddy didn't allow tantrums, so being able to control my emotions was not a challenge. I am often befuddled when I see white women in the workforce slamming doors and having emotional breakdowns, behavior that would without question send a Black woman home that very day. The perception of the "angry Black woman" is not easily dispelled. Keeping my composure was not always easy, but not impossible.

I bare the physical and emotional scars of being undervalued as a civilian serving alongside my uniformed counterparts. The scar on my right leg is a constant reminder of how little my service as a civilian to this country was valued. I fell down two decks when someone on the ship left a hatch open and forgot to put up the rails. The bone in my leg was visible from the fall for weeks and the wound was infected for months. I was still expected to report to duty with a smile on my face and figure out on my own how to care for my wounds, emotionally and physically. I suffered a miscarriage while surveying in the Gulf of Aden. The ship's captain lowered me into a rubber boat and sent me ashore to a "doctor" in Djibouti. I have no idea what medicines I was given before my mother wielded her power in the states and demanded that I be sent home immediately.

Civilians have been working under oath alongside their military counterparts in areas of civil unrest and in combat zones for many years in various capacities. Civilians served in the Vietnam War, the Korean War, both World Wars, The Mexican War and even in the United States Civil War. The demographics of civilians in combat zones has evolved since my experience facilitating the transition from the Cold War warfighting strategy to contemporary strategies. Since the 9-11 attacks, it is estimated that as many civilian personnel as uniformed troops are running U.S. military installations and posts in and around the world. Civilians in Iraq and Afghanistan provide security services to convoys and embassies. These plain-clothed civil servants and contractors maintain continuity, security, and provisions to ensure the mission to defend and protect the vital interests of the United States is possible.[51] Yet, those of us who have served in plain clothes in combat zones are not afforded any consideration of Veterans' benefits. At a minimum, Congress should consider allowing civilians who have served a pre-determined amount of time in combat zones some percentage of hiring preferences that are available to veterans.

Military strategy and civil service reform may ebb and flow with each administration and leadership personality, but the new is still the old. Outspoken restrictive nationalists' ideologies, fueled by an isolationist, intolerant, xenophobic, homophobic, sexist, and racist Trump administration are neither new nor are they fleeting. They are here to stay, further tainting both domestic and foreign policies. These ideologies that inform decision-makers contribute to the flawed democratic system in America and the failed attempts at democratization and peacekeeping led by the U.S. around the world.

There are those who would say that my experience was unique or that somehow my temperament hindered my progress. I would argue that within the corridors of America's workforce, you will find hundreds of stories with a common motif, especially among the demographic of Black employees who are nearing retirement. "*Sweet and docile, Meek, humble and kind*" is the key to longevity in the workplace. For those of us destined to "*uproot trees,*" our paths have been rocky and turbulent and there will be no celebration at the end of the journey. The hearts and minds of the elite are infused with generational misappropriation of ideas and practices that even in contemporary form serve as handmaidens of a disparaging state of affairs. It is this phenomenon that infiltrates and drives the political and socioeconomic fabric of America, which is run by direct descendants of ethnically-motivated ideologies passed down through generations.

As for me and my future, I will ride the last tides of life, theorizing, observing, and writing. As Daddy used to say, "Once you get it in your head, they can't take it from you." My head is full but my heart is still intact. I fulfilled my oath to *support and defend the Constitution of the United States against all enemies, foreign and domestic*. As far as a nautical chart for my career, I bid maritime strategy fair winds and following seas. I don't need permission to leave the ship. I've taken it. Unapologetically. Thanks for the memories. To the generation that follows, you have the watch.

FORWARD, FROM THE SEA

LESSONS LEARNED

I now know that from the very first day I swore to *support and defend the Constitution of the United States* that I created an identity crises for myself. I often wonder should I have fulfilled the oath I had taken before graduating from college to enlist in the United States Army rather than obviate that oath for a career as a federal government employee traveling around the world on U.S. Navy platforms. I also question whether I stayed too long, first with the Naval Oceanographic Office and then in the years that followed designing and building great ships and naval capabilities. When I left NAVOCEANO, I worked for two years with the National Weather Service and another two years at the National Institutes of Health. But the spirit in me assured me that I was a sailor and I belonged at sea. I grew up on the Gulf of Mexico and I spent nearly thirteen years sailing the world's oceans. There was a deep yearning that convinced me that I was a sailor too. Maybe I was, but the world that I worked in constantly reminded me that I was not. Even though I risked my life alongside military counterparts, I did not belong. I had not worn a military uniform. I was not a quitter, so I continued to bang my head against that wall of rejection. Naval officers who had never been in combat zones, never risked their lives, reminded me that I was not and would never be one of them. I accept that now.

Meritocracy in its present day form, the idea that people get what they deserve, what they've worked hard for is a myth, if anyone ever believed in that ideology anyway. In the federal government, everyone is not entitled to what they deserve, even the slackers. Hard work and education does not always payoff in the form of limitless opportunities. There will always be careless and often unconscious devaluation of race and insensitivity to culture in the federal government workplace, including in the government contractor space. Civil service reform has flip-flopped with each new administration, and no one has yet realized that the merit pay and promotion systems have never been equitable. The subjective nature of the system makes meritocracy impossible. Merit and mentorship must be supplemented with sponsorship. As evident by the constant on again off again appropriation of funds to open and close the federal government, civil servants are undervalued and dispensable. Partisan politics find the federal worker the easiest target for settling their disputes.

The notion that the solution to underrepresentation of minorities in STEM fields is to increase opportunities within the federal government is an illusion, a misguided attempt at a good faith effort of leveling the playing field. Sending highly intelligent, hard-working young people into an unfair and unequal work environment further exacerbates an already skewed work culture. Agencies are able to tout diversity successes at entry-levels, but those statistics are manipulated once employees start tapping on the glass ceiling. In a dynamic society that evolves with rapidly changing technology, young people should be encouraged to take multiple career paths. The idea that getting skills in one area is outdated. STEM students would be well-served to explore art and humanities fields while taking advantage of the opportunity to earn top dollars in science and technology careers.

Cultural illiteracy and apathy is partly to blame for failed policies at home as well as U.S. military and diplomatic strategy abroad, particularly in the Middle East and Africa. These failures can be attributed to proselytizing for a flawed American form of democracy around the world. Most accounts of that fateful day when a faction of armed Somali people shot down a UH-60 Black Hawk helicopter carrying U.S. Special Operations Forces in the notorious "Black Hawk Down" gloss over the fact that US and allied forced failed to understand Somali clan culture. The intelligence community admits there was a gap in "Human Intelligence" or HUMINT. Further, there is little mention in the media of the extraction of resources from the Somali people when Somali piracy is berated and the Somali people are unfairly portrayed as uncivilized. Even the Patriot Act which was intended to make it easier to track the financial activities of terrorist, has created a financial conundrum for countries where nomadic lifestyles are not conducive to traditional forms of banking. Cultural illiteracy and indifference has been a factor in Afghanistan, Iraq, Pakistan, and many other Middle East and African countries where military strategy has failed and abandoned the people to endure even greater atrocities than existed before the Americans arrived. As long as culture is minimized in U.S. policy and strategy, both foreign and domestic, the desired outcomes will continue to fail miserably. It is much easier for the U.S. to take a selfish approach—simply abandon the people to sort out the chaos created by external involvement.

Domestically, broad sweeping cultural literacy is necessary to mitigate widening polarization due to divisive rhetoric from the nation's leaders and mainstream media. The ease at which fake news, photos, and videos are infiltrating social media constructively facilitates cultural divisiveness. It is important that the digital generation can distinguish fact from fiction when it concerns dehumanizing "the other."

Finally, I have resolved that fair and equal is not achievable, but being true to one's self and having faith in your own ability is possible. Know when to say "when." There are multiple paths to achieving one's personal best and each one of them should be explored. As my Daddy use to say, "You're J.B. Carter's daughter" and that's enough to live up to.

ACKNOWLEDGMENTS

An infinitely wise God blessed me with parents, J.B. and Mary Carter, who came together and formed just the right mix of fearlessness and strength for me to believe I could be anything I wanted to be and do anything I wanted to do. I am grateful. My siblings and I bear the integrity of the family name as well as the initials my parents thought should be carried on forever. To my brother J.B., III and my sister, Janelle Betrice: I, Joy Bonita, thank you for enduring my forever evolving personality as the middle child.

Imani. If I could redo the years I spent traveling around the world, leaving you to be super strong and independent, I would take them back and make you my little girl all over again. You are my reason for living and pushing and never giving up. I am so proud of the woman you have become.

Marcel, "Cake." You kinda understand that I am a restless soul. Thank you for not trying to change who God made me to be.

Shanda. You were my first daughter. I could not have asked for a better bonus daughter. Sweetness is your middle name.

Marcel, "Two." You brought the sunshine. You will understand one day all the light I see in you.

Many people supported me in writing this memoir, taking the time to read sections of it and providing helpful comments; To my sisters of Delta Sigma Theta Sorority Incorporated, especially Connie, Yvette, and Quan: thank you for your time and attention to reading the manuscript. To my sister, Janelle who jumped in to rescue me from a state of confusion when I needed structure the most, Muah! To my sister-in-law, Veronica who sat for hours patiently walking through my manuscript, thanks. To my friend and colleague, Terry Y. thank you. And finally, to my pastor and friend, THE Dr. James Henry Harris, I know you think I don't listen, but I do. Thanks for your time and patience.

REFERENCES

1. Venter, Al J. *Somalia: Unending Turmoil, Since 1975*. Pen and Sword Books. 2017
2. Mississippi Department of Archives and History. http://www.mdah.ms.gov/arrec/digital_archives/sovcom/
3. http://americanradioworks.publicradio.org/features/mississippi/f1.html
4. https://www.youtube.com/watch?v=CPV6ubWDun0
5. Dr. James Henry Harris
6. http://www.public.navy.mil/FLTFOR/cnmoc/Pages/History_NAVO.aspx
7. Day, John Kyle. *The Southern Manifesto: Massive Resistance and the Fight to Preserve Segregation* 2015
8. Curtis, Jesse N. Remembering Racial Progress, Forgetting White Resistance. History & Memory, Vol. 29, No. 1 (Spring/Summer 2017)
9. Tumulty, Karen. Trent Lott's Segregationist College Days. Dec. 12, 2002. http://content.time.com/time/nation/article/0,8599,399310,00.html
10. https://larouchepub.com/eiw/public/1996/eirv23n25-19960614/eirv23n25-19960614_075-what_does_the_senator_from_missi.pdf

[11] https://www.washingtonpost.com/archive/lifestyle/1994/04/25/sen-lott-and-the-nasa-contractor/177a42ac-5185-463d-b637-46a7b2f42f10/?utm_term=.b4b7c14350c6

[12] Prince George's County, Maryland v. Holloway, 404 F. Supp. 1181 (D.D.C. 1975); https://law.justia.com/cases/federal/district-courts/FSupp/404/1181/2131971/

[13] http://www.usdiplomacy.org/history/service/jamesbishop.php

[14] https://www.washingtonpost.com/archive/lifestyle/2001/04/29/confessions-of-a-hero/a34756ad-6f80-4154-a890-504e9267a8a0/?noredirect=on&utm_term=.197943ccf373

[15] National Military Strategy of the United States https://history.defense.gov/Portals/70/Documents/nms/nms1992.pdf?ver=2014-06-25-123420-723

[16] http://archive.defense.gov/news/newsarticle.aspx?id=41585

[17] Highlights of GAO-04-380, a report to the Ranking Minority Member, Committee on Commerce, Science, and Transportation, U.S. Senate

[18] http://www.ndia.org/about

[19] https://www.sba.gov/blogs/sbas-8a-certification-program-explained

[20] http://www.seaport.navy.mil/

[21] SBA's Historically Underutilized Business Zone Program designed to help small businesses in rural and urban secure government contracts. https://www.sba.gov/blogs/sbas-hubzone-program-helping-urban-and-rural-based-businesses-succeed

[22] https://www.sba.gov/offices/headquarters/ogc_and_bd/resources/11498

[23] https://washingtontechnology.com/Articles/2009/08/31/Cover-Small-biz-Top-ANC-chart.aspx

[24] https://www.census.gov/content/dam/Census/library/publications/2017/acs/acsbr16-01.pdf

[25] https://www.fedsmith.com/2015/05/20/fedsmith-com-users-think-racism-is-a-problem-in-the-federal-workplace/

[26] Navy Irregular Warfare Vision for Confronting Irregular Challenges

[27] https://link.govexec.com/view/5a6d89a866c37951618ba-22b8iiv9.5iz/eeb89f44

[28] https://sites.tufts.edu/corruptarmsdeals/2017/05/05/the-boeing-tanker-case/

[29] https://sites.tufts.edu/corruptarmsdeals/2017/05/05/the-boeing-tanker-case/

[30] http://archive.defense.gov/pubs/ASB-ConceptImplementation-Summary-May-2013.pdf

[31] https://icrd.org/

[32] Cox, Brian. *Faith-Based Reconciliation.* 2007.

[33] Espinosa, Gaston. "Barack Obama's Political Theology: Pluralism, Deliberative Democracy, and the Christian Faith." Political Theology 13, no. 5, 2012: 610-633.

[34] https://www.nap.edu/read/1751/chapter/4

[35] The Loyalty Review Board of the U.S. Civil Service Commission, 1947-1953 Author(s): Henry L. Shattuck Source: Proceedings of the Massachusetts Historical Society, Third Series, Vol. 78 (1966), pp. 63-80 Published by: Massachusetts Historical Society

[36] Ingraham, Patricia W., Rosenbloom, David H. *The Promise and Paradox of Civil Service Reform.* 1992 University of Pittsburg Press.

[37] Ibid 32

[38] http://yalejreg.com/nc/president-george-h-w-bush-on-civil-service-and-civil-society/

[39] J. Edward Kellough , Lloyd G. Nigro , and Gene A. Brewer ; Civil Service Reform Under George W. Bush: Ideology, Politics, and Public Personnel Administration

[40] Wagner, Erich. *Cautionary Tales from Past Attempts at Pay-for-Performance.* www.govexec.com 08/16/2018 https://www.govexec.com/management/2018/08/cautionary-tales-from-past-attempts-pay-for-performance/150615/ Accessed 08/18/2018

[41] https://www.fedweek.com/fedweek/time-growing-short-to-reform-the-reform-act-this-year/

[42] https://www.cnn.com/2018/08/30/politics/trump-cancels-federal-employee-pay-raises/index.html

[43] Office of Personnel Management. https://www.opm.gov/fevs/about/

[44] http://www.navy.mil/navydata/people/cno/Richardson/Resource/Navy_Civilian_Framework.pdf

[45] https://federalnewsradio.com/workforce/2018/05/who-they-are-and-where-they-work-the-federal-workforce-by-the-numbers/slide/1/

[46] Bonikowski, Bart, and Paul DiMaggio. 2016. "Varieties of American Popular Nationalism." American Sociological Review 949–980.

[47] dhs.gov. 2017. *Fact Sheet: Executive Order: Border Security and Immigration Enforcement Improvements.* February 21. https://www.dhs.gov/news/2017/02/21/fact-sheet-executive-order-border-security-and-immigration-enforcement-improvements.

[48] bbcnews. 2017. *Trump's executive order: Who does travel ban affect?* February 10. http://www.bbc.com/news/world-us-canada-38781302.

[49] No Longer Bound

[50] Rusin, Jo B. Move to the Front: A Guide to Success for the Working Woman. 2001. Presideo Press

[51] https://www.your-poc.com/u-s-military-used-civilian-contractors-combat-zones-since-1960s/